The Sustainable Project Manager: Constructing the Water Systems of the Future

The Sustainable Project Manager: Constructing the Water Systems of the Future

Pernille Ingildsen and Mette Thorkilsen

IWA PUBLISHING

Published by **IWA Publishing**
Unit 104–105, Export Building
1 Clove Crescent
London E14 2BA, UK
Telephone: +44 (0)20 7654 5500
Fax: +44 (0)20 7654 5555
Email: publications@iwap.co.uk
Web: www.iwapublishing.com

First published 2024

British Library Cataloguing in Publication Data
A CIP catalogue record for this book is available from the British Library

ISBN: 9781789063332 (paperback)
ISBN: 9781789063349 (eBook)
ISBN: 9781789063356 (ePub)

Doi: 10.2166/9781789063349

Dedication

This book presents decades of our research and personal experiences from working on projects as well as the experiences of project managers we have worked with in the role of consultant, manager, steering group member, or member of the team. We regard ourselves as professionals who view project management as a tool for achieving goals, rather than the goal itself. However, because project management is such a crucial vehicle for change and creating value, we have dedicated a lot of attention, reflection, and thought to understanding how to succeed with project management. We have conducted hundreds of experiments and adjustments to improve our own project management. The projects and the task of project management have been enriched by people who have inspired us.

We dedicate this book to all those who have crossed our path in project management: employees, trainees, managers, coaches, trainers, and peers. Thank you for the inspiration.

We also dedicate the book to the sustainable project managers of the future. Wishing you an interesting journey and the best of luck.

Contents

doi: 10.2166/9781789063349_xi

Preface by Professor Gustaf Olsson

This book is more than a management consultant book. Taking the message from the book seriously means much more than learning several management rules; I ought to reconsider my own attitudes, how I deal with people, and how I respect the planetary boundaries. The book describes several aspects of management: management as a discipline, my own attitudes, and my interaction with others.

I have had the privilege to work closely with Pernille Ingildsen for more around 25 years. She is a unique combination of an internationally recognised researcher and engineer, scientific editor, author, philosopher, and manager, having a deep understanding of many aspects of what a meaningful life should contain. Her cooperation with Mette Thorkilsen makes up a most thought-provoking team. Mette has a considerable experience on leadership in organisations. She has researched and practiced organisational systems and has provided deep understanding what is required of a new mindset leading to a servant and personal leadership that is intelligent, beautiful, and fair. Mette has described how such an organisation should be transformed, from a command-and-control system to an interest-based system.

The word 'sustainable' in the title is given a meaningful and multifaceted interpretation. I am sufficiently old to vividly remember the 1968 Christmas Eve, when the Apollo 8 crew shot the picture that became an icon. It showed the rising Earth, the blue dot in the endless space: *this is all we have.* I can still recall the awe and wonder of seeing our vulnerable home, a picture that would forever change our perspective of our limited world. Five years later, the Oxford economist Ernst F. Schumacher (1911–1977) published his landmark book *Small is Beautiful*. He talked about 'natural capital' and chose fossil fuels as one example. We are treating them as income although they are undeniably capital items: 'If we treated them as capital items, we should be concerned with conservation: we should do everything in our power to try and minimize their current rate of use; … We are not in the least concerned with conservation; we are maximising, instead of minimising, the current rates of use'. Schumacher warned that the modern world must balance economic growth with the human costs of globalisation.

Schumacher mentioned the other aspect of our limited resources, the tolerance margins of nature: water resources and water scarcity, polluted waters, dams from mining operations, limited resources of vital substances like phosphorus, and waters and wetlands that have been devastated by oil operations. He seriously questioned the idea of unlimited growth, 'more and more until everybody is saturated with wealth'. We know that the environment will not be able to cope with the burden of growth. Modern economy is driven by greed and envy and Schumacher claimed that the inevitable result is 'nothing else than a collapse of intelligence'. We, humans, threaten the beauty of the Earth with our overexploitation of all that can give us money and power. If we understand that we are a part of nature, it will become clearer that a battle with nature is a war against humanity.

Having talked about *silent spring, small is beautiful,* and *limits to growth* during the 1960s and 1970s the term sustainability could express the general goal about the ability of people to co-exist on Earth over a long time. The term became much more recognised after 1987, when the United Nations (UN) Brundtland Commission defined sustainability. Today we have got a deeper insight into the planetary boundaries, and we have expressed and accepted the sustainable development goals (SDGs). Also, a radical change of economics is needed, like the doughnut economics thinking. For example, if we would accept national capital accounting of rainforest as a basis for the price, it would be considerably more expensive to eat meat.

Pernille and Mette show us that sustainability thinking is a fundamental attitude to life, to economy, to consumption, and to lifestyle, as well as to managing projects. It is a guiding rule that will force us to consistently test our values, not only to natural resources and economy. Also, people – project members – should be sustainable. We are also reminded that we all have a personal responsibility. The authors emphasise that the overarching responsibility of the project manager is to take the very pragmatic approach: *what can I do*? This requires a real in-depth understanding of what sustainability is all about, and the ability to think both holistically, and long term. This competence is seldom encouraged in our quarterly financials thinking. The authors remind us about the planetary boundaries and encourage and inspire us to think and reflect how the project leader role fits into the larger picture.

In the first part of the book the authors describe fundamental management knowledge for decision makers as well as challenges that almost always will appear during the project phases. Given the spirit explained in the first part of the book it is obvious that good decision making needs to have a suitable structure. It must contain a value system that is based on long-term thinking. Naturally, it is easier to make competent decisions in a small project than in a complex one. However, the challenges are similar: try to understand the project and all its constraints. The authors emphasise what is increasingly ignored in our time: take responsibility, even when we don't know everything. Communications skill is a critical ingredient, not the least in the balance act between time, resources, and quality. It also includes the communication between the project owner and the project manager. Mutual trust is fundamental for any organisation.

Pernille and Mette will reiterate that *wisdom* is an important ingredient in good decision making. Being *smart* will take the constraints of time, resources, and quality into consideration. Being wise incorporates a broader set of capabilities that address the full scope of decision making, including both humans and nature. Giving co-workers the environment and support they need and trusting them to get the job done is part of the wisdom. Again, we the readers are forced to reflect at the end of each chapter so that we do understand the messages.

Five basic management tools are explained. They are meant to meet the challenges presented: clear objectives, trust between project owner and project manager, responsibilities, and balance between time, resources, and quality, engagement. The five tools – objectives, stakeholders, milestones, project organisation, and risk analysis – have been tried and tested over many years and are presented here so pedagogically. We can learn from mistakes and poor planning, and it is crucial to do the homework. Poor implementation of any of these tools is to ask for small or large failures. How do we know that we have succeeded? Pernille and Mette help us to grasp the task. A thorough work in planning forces us to understand the project. It is so important trying to predict potential risks and difficulties before they will appear. We all know that a world without disturbances does not exist, so good management ought to be able to respond adequately to troubles, turbulences, and conflicts. As a reader I am encouraged to reflect to prevent that my knowledge will remain passive.

The concept of *feedback* is crucial for the successful manager, which is highlighted several times in the rest of the book; act and respond to real disturbances. Similarly, it is emphasised how prediction is part of successful management. Good prediction also leads to early actions to avoid or counteract troubles. This is the key to feedforward actions.

The authors emphasise that the manager of the future must systematically work to include sustainability in the role. This requires that we change ourselves and not only repeat what we have always done. This is difficult for anybody. Still, it is necessary and requires a lot of self-discipline. Like any attempt to change the world, we must begin with ourselves. Then the rest of the world may follow. The awareness of sustainability makes us realise that the world is changing rapidly, and the demand for sustainability is no longer as controversial as it used to be, but forces any manager to consider sustainability, time, people, and economy, combined. We can no longer defend poor project management by claiming that 'we have always done like this'. Pernille and Mette make it so clear that the whole mindset should be open to manage all competing requirements to meet the goals of a complex project.

In our time, when specialisation has evolved dramatically, it is increasingly important that some people are a combination of generalists and specialists. The manager should act like the conductor of a symphonic orchestra, knowing sufficiently about all the parts in the musical score, but not necessarily a master player of all the individual instruments. Still, if the conductor is very good playing one instrument, he or she will be respected by the orchestra. The conductor must also know sufficiently about each of the instruments so that he/she can guide the musician how to improve the interpretation, balance the individual instrumental contributions, thus obtaining the perfect orchestral sound. The interpretation – making the music to touch the listener – makes the difference between the great and the mediocre conductor. The future manager should try hard to achieve this art of interpretation using 'leadership from the heart triangle'. In other words, do not forget the five basic skills, but build on them. Every manager needs to raise the competence, and the authors demonstrate how this can be measured. The wise manager will keep a distance between incident and reaction to allow the necessary reflection. The successful project group or a larger organisation depends strongly on mutual trust, cooperation, and empathy. The manager is the key individual to make this a reality. Too often we look in vain for an ethical compass. So, how do we become the builders of a sustainable future?

At this point in the book Pernille and Mette will guide us how to lead ourselves, becoming wise managers. They emphasise that we should 'get a sense of ourselves' and dare to ask the difficult and essential questions who we are, what we want and why. Management is so much more than a book of recipes and rules. Ethics does not seem to be a main priority in our time, but the future needs it more than ever. Truly, so many tasks are difficult, complicated, and overwhelming, but we should look for ways to act. Stoicism is one of them, teaching us to concentrate on things that we can do something about. This is not novel management; it is a philosophy and knowledge developed during millennia. We better learn from our ancestors. We also get acquainted with Daniel Kahneman's pioneering book and thinking: the dilemma of 'thinking fast or slow'.

Our 'inner work' is an indispensable part of solving complex problems. It boils down to the ultimate challenge how the heart can support the brain and how the brain can support the heart. Another way to look at it is that the end and the means are equally important. Most projects of any complexity are much more than technology, so we need to understand elements of human behaviour and what lies behind it. Having this insight, we may be able to adjust behaviour to achieve the desired intention. Use our eyes and ears!

Again, Pernille and Mette force me as a reader to do more than quickly read the chapters. I must reflect and translate the text to my own person and my own professional task. How is my attitude to the team members? Do I listen more than I talk? Do I try to look at a problem also from the other person's perspective? Am I pro-active? Do I keep the goal in my mind when making my decisions along the journey? Do I distinguish between urgent and important? Do I have confidence in my role? Do I possess the inner discipline that is required? How do I establish a process of change that produces results – for me and for the project? Patience – which is not the same as working slowly – is fundamental. We can learn how to lead ourselves.

Now comes the challenge to interact with others. How do we succeed in making two plus two more than four? Normally the project manager does not have the power to command and control. More often the role is based on trust, mutual respect, team roles, and common interest. The true manager task is to build the team in such a way that everybody can contribute. We can learn from the nine team roles. They look different during the phases of the project. Again, Pernille and Mette remind us that trust is a key ingredient, always, and emphasise how trust is built on various components. It is no surprise that conflicts may appear in any team and at any project phase. Then we should be prepared to approach the conflicts in a constructive way. There is a lot to be learnt in handling conflicts, and the authors encourage us to reflect.

The construction of the Øresund Bridge in the 1990s is cited as a role model for excellent partnering and management in a truly complex project. I happened to live on the Swedish side of the bridge, in Malmö, and could almost on a day-to-day basis, looking out from our apartment, see how the connection between the two countries was getting completed. A lot of lessons could be learnt from the project, and the right partners can contribute to truly win–win situations. Still the same lesson: try to build mutual trust! If we can decide long term, think strategically, and try more 'impossible' solutions, then we have come a long way. Never forget the goals of sustainability, co-workers' satisfaction, and cooperation are as important components in a project as time, economy, and quality. We should recognise that only when we find ways to change ourselves and our patterns of thought and behaviour we can hope to succeed in leading others through transformations. If we can fully understand and apply the lessons from this book, then we have come a long way in our attempts at creating the future's sustainable society.

Gustaf Olsson

doi: 10.2166/9781789063349_0001

Introduction

We are called to be architects of the future, not its victims.

Buckminster Fuller

Time and time again, the skill of leadership is recognised as the most important competence for success as a project manager. However, again and again, project management books and courses are mostly about tools. Tools are important, and without them, projects won't get off the ground. However, once you have learnt the most important tools, you ought to shift your focus and attention to developing your ability to lead. The impact you can create through projects depends very much on who you are and how you manage. We believe that being a sustainable project manager is more about working style and being on a developmental journey, than rigidly adhering to predetermined process descriptions.

We (the authors) have worked with project management and sustainability for decades. Mette has served as a project manager on major projects such as the Øresund Fixed Link (tunnel/bridge connection between Denmark and Sweden), the new headquarters of the Danish public service broadcasting company (DR), and as a technical manager for low-cost housing. Additionally, she has worked as a trainer in project management and portfolios. Pernille, on the contrary, has held roles as project manager and manager of project managers in water and wastewater utilities, development departments, and academia.

One day, we had an intriguing conversation about what it takes to become a sustainable project manager and the meaning of sustainable project management. We wondered when we would reach the point of being exactly where we need to be and how we can get there? After 2 years of writing this book, we can now answer this question with some confidence. In our understanding, sustainable project management consists mainly of the following four key ingredients:

- Understand sustainability, sustainability models, and align your practice accordingly.
- Apply the project management tools explicitly and pro-actively.

- Learn to lead yourself and keep on with your personal growth process.
- Design and implement processes that create trusting working relationships.

This book is written for all project managers who are motivated by sustainability and who want to take on the task of co-creating a sustainable world, create good and robust projects in their field, and succeed with developing themselves.

Before we get started for real, try to imagine the best sustainable future. Imagine a world where you can sense that the world feels good. You are safe, you take responsibility, you are healthy and strong, and you can express your creativity. People around you are happy – and you yourself are naturally happy when others succeed. You feel competent with the people you work with, and you're constantly finding new ways to make projects even better – and the results you create are like little gifts for the future.

That's what the idea of sustainability is trying to create, and every grand or small step helps. Every sustainable project moves us in the right direction and expands our sustainable capacity. When enough of us join this extraordinary movement, transformative change can be achieved. So, feel it! Look at what you are doing right now, in the light of everything you know: does it feel good? What modifications can be made to promote a better, healthier future?

Sustainability is about responsibility, generosity, and faith in the future. Faith that one day someone will look back at us and think that we were good ancestors to them. As project managers, we are always creating the future. It's difficult to grasp the severity of the sustainability crises. It's also challenging to implement new effective project management techniques. It's demanding to create change both in ourselves and in the environment, we live in. But moving forward we are all going to face these kinds of challenges again and again. It is natural to have doubts about what course of action is best. Doubt is our friend. If we feel overconfident, it might be because we are copying our shared past – and that's not good enough anymore. So, prepare yourself that this requires new learning, and that any new learning requires openness, curiosity, and a willingness to doubt. The worst that can happen is if we turn ourselves and our gaze away or continue with business as usual.

Instead, we need to be dedicated and disciplined in our development. Keep adjusting everything. Every difficulty is an opportunity to come up with a better and more creative answer. Keep going! Climb that mountain, run that marathon, build that house, create that garden – keep going!

This book is based primarily on experiences from the water sector and the construction industry, but project management competence is very generic, therefore this book can also be used and inspire other sectors. On the surface, water sector and construction projects do not differ significantly from other complicated projects, but that said, the industries have their own challenging characteristics that are worth paying attention to. Water and construction projects are characterised by products that have a very long durability and lifespan, they involve heavy and not very flexible systems – and compared to, for example, the IT industry – these projects are not very open to iteration once construction is underway. Think of the times when water construction systems don't work. These constructions are easily some of the most critical infrastructures that we have. Malfunctions have wide and heavy consequences on our society and environment.

Given this importance of water infrastructure – in terms of its criticality for a well-functioning society in a healthy environment not only now, but also in the future – we believe that project management has had way too little attention in the water sector. Project management is not difficult per se, but it is important that it is carried out carefully and consistently – otherwise project management becomes increasingly difficult and costly as a project progresses ungoverned. We need to have project managers who think critically, acts proactively, and have the future of sustainability in mind at all times. Therefore, we have written this letter for you from heart-to-heart.

Dear Project Manager

We are in the process of transitioning from an industrial society to a sustainable one, although we haven't fully reached that point yet. It is crucial for us to make this transition as soon as possible. The success or failure of this endeavour, to a large extent, relies on you and your projects. You have literally the power to design and construct the future.

In the upcoming period, many aspects will transform, presenting new challenges for you and your project management. What was considered normal yesterday may no longer be the norm tomorrow, and what was deemed satisfactory yesterday may not suffice tomorrow.

We don't have all the answers to what sustainable project management will look like, but we do have years of experience as project managers. By utilising this resource, you can achieve success more efficiently and effectively than attempting to navigate the complex world of sustainability on your own. To succeed, particularly four areas require your focus:

In-depth understanding of sustainability

The news is full of stories about sustainability, and the majority of people watch the news from the sidelines. But when faced with the responsibility of project management in the midst of a transformational time, it is necessary to have a significantly deeper understanding of the issue than a spectator. It is a complex field with many interdisciplinary specialities. Many of the problems are systemic in nature and cannot be solved at the snap of fingers. They require difficult balancing and attention – often balancing conflicting considerations. If you do not continuously monitor and build up your own internal picture of how it all fits together and what your role is, you run the risk of spending your energy in the wrong place. It is therefore necessary to familiarise yourself with planetary boundaries, the UN sustainable development goals and different models for sustainability and systems analysis. Keeping up and keep learning about the meaning and evolution of our common understanding of sustainability is a continuous endeavour. It requires both water knowledge, but also a broader understanding of the philosophy of sustainability.

Project management needs to be fully balanced

Project management has been an independent discipline with textbooks and formulated theories since the 1950s and has evolved a lot in the last 70 years (Olsson *et al.*, 2019). However, the discipline is still not common practice everywhere and something that people keep skipping too lightly over. Hence, the basic tools and mindsets are still only partially implemented in most places. Too many project managers are confused about the use of the tools and cannot confidently stand up in project teams, steering groups, and among stakeholders and explain how their project is managed. Too many projects stumble in basic understanding of purpose, success criteria, deliverables, schedules, budgets, risk analysis, stakeholder management, and organisation. Now is the time to put these basic skills into your core practice and be very specific and precise about it. It is not difficult it just needs to be done consistently.

Your connection to yourself is important

Currently, there is significant tension between individuals who wish to preserve the status quo and those striving to adapt to the demands of our changing world. These discussions are an expression of resistance to change and are as such 'natural' discussions to have. As a project manager, it is crucial that you work consistently on being aware of who you are, what your values are, how you contribute, and how you work with other people. A personal position that is built on thinking independently and being knowledge-based ensures that you can navigate these conflicts. Project management is a leadership job! And project leadership will be even more important in the coming years than traditional management. The future is created through projects. Therefore, project managers must take leadership upon themselves. Leading others requires self-awareness and strong connection to one's true self.

Your ability to facilitate co-creation

The ability to collaborate has been a core winning competence for most of human history. In recent years and decades, new research and theories have emerged on how to make collaboration work quickly and deliver exceptional results. At the heart of the project manager's role is the task of making collaboration work. What is new is that this collaboration should not simply replicate the same results as in the past but should be able to deliver results based on society's needs in a sustainable future. This means that everyone in the team needs to get out of their comfort zone and into the competence development zone. Therefore, the project manager must learn to become able to create a safe space to work creatively as well as learn to facilitate the creative processes. The future is collaboration, partnerships, and co-creation. Your projects and project results are your legacy for the future.

Ultimately, it's about all of us, but not in an anonymous collective way where no one really steps up – on the contrary, it requires a future where we all step up – and that starts with you and me.

We hope you enjoy the book and wish you good luck on your journey.
Best regards
Pernille Ingildsen and Mette Thorkilsen

Sustainability

doi: 10.2166/9781789063349_0009

Sustainability: the problem and the opportunity

Our biggest challenge in this new century is to take an idea that seems abstract – sustainable development – and turn it into a reality for all the world's people.

Kofi Annan

Having an in-depth understanding of sustainability and the pressing sustainability crises is central to acting sustainably in projects. Anyone who wants to work as a sustainable project manager needs to understand the fundamental dynamics of sustainability. This chapter provides you with a foundation consisting of a series of unified models to help you understand the complexity and bring some order to it.

The understanding that came with the first images of the Earth from the outside has been central to our understanding of the crux of our crisis. Suddenly, we could see and imagine ourselves as one entity seen from above. This 'above-view' understanding is the basis for recognising the planetary boundaries within which we need to stay if the conditions for life on Earth are to be liveable for the biosphere that we, as humans, are part of and thus dependent on. We should give the same attention to the health of our biosphere as we do to the health of our bodies – at this point perhaps even more so. If we treated our body as we treat our biosphere, we would be in deep crisis. And the theory of the nine planetary boundaries is a story of crisis. There are nine planetary boundaries that need to be understood by heart. The trespassing or near-trespassing of these boundaries constitutes the root causes of the many spin-off crises that creates situations that results in critical situations and on the long run are unsustainable. When we exceed planetary boundaries, it is not possible to maintain the long-term health of the biosphere. The gradual degradation has effects that escalate over time and interact with each other. This biosphere is us; it is other species; and it is our home. Basically, what is very clear from the 'above-view' of Earth is that all living beings of this biosphere are interdependent on each other.

The concept of sustainability is easy to understand as a general concept. Sustainability is the ability to maintain or preserve something over time. In the context of the environment and natural resources, it can be understood by Gro Harlem Brundtland's definition (Brundtland, 1987), where sustainability defines a practice of using resources in a way that meets the needs of the present without compromising the ability of future generations to meet their needs. It involves finding ways to balance economic, social, and environmental considerations to create a just and habitable world for all (Brundtland, 1987).

The principle of sustainability can be applied to many different arenas and practices. Sustainability is emerging in more and more areas, perhaps most visibly in environmental protection, resource management, development, social justice, and public health. But it is also possible to talk about sustainable personal practices, building and construction practices, project management practices, and so on. There are several principles that follow from the idea of sustainability that are applicable across the board. Understanding sustainability requires a holistic approach that considers the interconnectedness of everything, but also to be practical a more direct approach addressing key factors. Both the whole view and the partial view should aim at improving the long-term impacts of our actions. Ultimately, the goal of sustainability is to create a world that can meet the needs of all people, now and in the future, while preserving the natural environment and its resources so that they are available for future generations.

But the 'simplicity of sustainability' can easily seduce. Currently, many of our practices are unsustainable, and it has proven challenging to alter the course of how we have structured our lives and societies. Consequently, we now face a multitude of pressing sustainability crises from various angles – at the same time as we need to change course.

A major obstacle to securing a sustainable future is the way we approach economics. Economics is such a strong force for us that it sometimes seems as if it is almost like natural laws governing our actions. This is however not correct. Economics are game rules that we as humans have invented and implemented – and as such, they can be changed – it's just complicated. Economist Kate Raworth has thought carefully about this. She offers a professional critique of the way we work and think about economics and has proposed a different way forward (Raworth, 2018). Understanding Kate Raworth's doughnut model is a prerequisite for understanding which social considerations need to be balanced to successfully steer our society back within the planetary boundaries.

Raworth's doughnut economy thinking leads us naturally to the most ambitious governance endeavour of our time: the SDGs. Adopted in 2015 at the UN, SDGs are a huge global leadership experiment: how can we synchronise the whole world to steer towards the same goals without being able to force anyone? SDGs are a shared vision of the future agreed by the countries of the world. It is up to everybody to achieve them.

SDGs are defined in a global way and therefore it may be difficult to translate it to our everyday context. For the water sector a primary way of translating sustainability is looking into water stewardship as means of focusing on water. For the execution phase of many water infrastructure projects, it is relevant to look at building and construction certifications for sustainability. Different certifications are available in different countries for translation of the SDGs into building practices.

Familiarising oneself with these different layers of the understanding of sustainability helps to understand the challenge and thus enable oneself to act wisely. In addition to the above-mentioned frameworks for understanding, there are many other models and concepts that may be relevant in different contexts. These models and concepts help to create 'thinking and calculation tools' to navigate the complex world of sustainability. A few of the very promising concepts are shortly described as well.

The high level of frustration in society about the slow progress is completely understandable. Deep understanding provides an insight into why it is slow; it is complicated. Complicated – but not impossible.

THE PLANETARY BOUNDARIES

The clarification of the planetary boundaries has made it possible to monitor the development in the field of sustainability at a global/planetary level. Most actions are local and thus need to be aligned and directed towards tackling the major issues. By understanding the nine planetary boundaries and monitoring the direction in which the Earth's overall parameters are moving, it is possible to get an overview of the consequences of the billions of actions that lead to the major systemic and global effects. A significant aspect of problem-solving involves making informed decisions and collaboratively executing the required initiatives and projects. But as in all good project management, it is important to understand the purpose before we plunge headlong into the solution. Therefore, read this chapter for an overview of the sustainability problem.

The Earth is about 4.54 billion years old. The first life to emerge was single-celled life, but after photosynthetic production of oxygen in the atmosphere emerged among these ancient organisms, the conditions for multicellular life were created. The first complex multicellular organism appeared 540 million years ago. If the entire Earth's history is spread over a 24-h period, this life emerged at around 21:50. This was followed by a long period with a temperature 5–10°C warmer than today and a water level 70 m higher than today, which would have made human life impossible. This 'hothouse' period was followed by unstable conditions, with successive ice ages rolling over the Earth. As carbon dioxide (CO_2) levels fell, the Earth entered a stable period, which has now lasted for 12 000 years. In terms of a human lifetime, that's a very long time, but in terms of Earth's history in 24 h, the 12 000 years correspond to less than a second before midnight. So, this stability that we take for granted is anything but. With the increased CO_2 concentration in the atmosphere, we may be coming to the end of this stable period. This change is so significant that we are entering another geological age, the anthropogenic or Anthropocene age – a geological age named after the effect of human beings.

In the book 'Breaking Boundaries – The Science of Our Planet' (Gaffney, 2021), this time is compared to driving a car on a steep downhill road with hairpin bends and no guardrails at night and with the headlights off. Instead of slowing down the car, we are accelerating. This drive represents the route out of the peaceful 12 000-year Holocene age towards a new unknown state. What that age will be like depends on the ability of humans – all of us – to navigate the planet based on an understanding of 'Planet Stewardship'. It is not without reason that the children in the back seat are protesting. The difficulty is that the transformation from the Holocene conditions for life that we know, to a different future state is hard to see from an ordinary lived life. It is outside our frame of experience and the change is slow in human time, but extremely fast in planetary time. The only way we can get an overview of the situation is through scientific knowledge and measurements. Science says – pretty much with one mouth – that we must actively (re)establish a state on Earth that is within a set of planetary boundaries that ensure that the Earth continues to make it possible for humans to live here. This is a difficult insight to live with, and it creates disbelief in society. We simply do not have a similar experience from the past.

Coronavirus disease-2019 (COVID-19) gave us a test period of this way of thinking. In Italy, which was the first country in Europe to be hit by Corona, physicist and mathematician Paolo Giordano wrote as early as in the spring of 2020 the book 'How Contagion Works – Science, Awareness and Community in Times of Global Crisis' (Giordano, 2020). Although the sustainability crisis moves slowly in terms of human time, the Corona crisis had a speed that made it easier and clearer to see the mechanisms that have taken a long time to realise in terms of sustainability. Hence, there are lessons to be learned from the COVID-19 crisis.

Paolo Giordano writes (Giordano, 2020):

> And so, the epidemic encourages us to think of ourselves as belonging to a collective. It pushes us to behave in a way that is unthinkable under normal circumstances, to recognize that we are inextricably connected to other people, to consider their existence and well-being in our individual choices. [...] If there are many of us, each of our choices has global consequences, sometimes abstract and hard to understand: during a contagion, the lack of solidarity is first of all a lack of imagination.

Similarly, project management in the age of sustainability is different because we can now see the global responsibility that lies on the whole project execution, that is, in terms of use of materials, use of technologies, use of energy, and so on. And we can easily long for a bygone era of less complexity – but going back is not possible. We have opened our eyes to the understanding that we are part of a great planetary whole in crisis – and we should be careful not to close our eyes again to that fact.

When we view life through the planetary boundaries the Earth appears as a spaceship and the boundaries tell us the same as a pilot's training, 'if you don't keep the aircraft's parameters within these boundaries, the plane will crash'. The planetary boundaries constitute a simplification, but an effective one. It prescribes that for sustaining a planet with an environment conducive to our civilisation, there are nine crucial boundaries or limits that must be adhered to:

(1) Climate change
(2) Ozone loss

(3) Ocean acidification
(4) Loss of biodiversity
(5) Use of land
(6) Fresh water
(7) Biogeochemistry
(8) New substances.
(9) Atmospheric aerosols

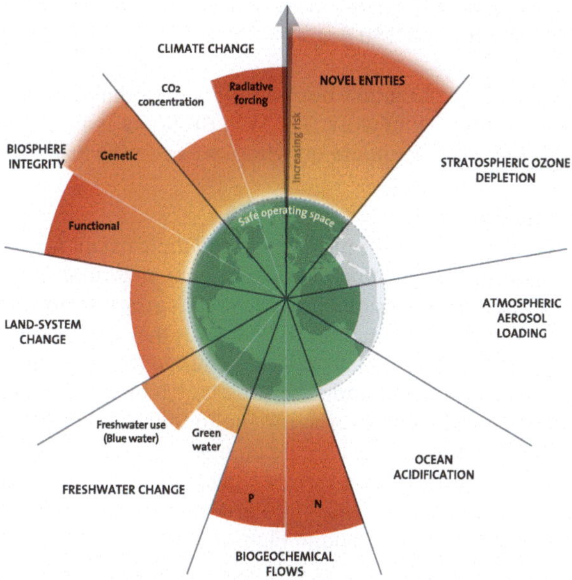

Figure 1 Status of the nine planetary boundaries 2023 (*credit*: Azote for Stockholm Resilience Centre, Stockholm University. Based on Richardson *et al.* (2023), Steffen *et al.* (2015), and Rockström *et al.* (2009).

Each of the nine boundaries has its own complexity. The nine planetary boundaries are divided into three groups representing three systems:

(1) Climate change, ozone layer, and ocean acidification are large global systems where the whole planet is affected in a simultaneous and immediately distributed fashion. In this sense, they are the 'Big Three' planetary boundaries where only global cooperation will be helpful.
(2) The second group is the four boundaries that deal with the biosphere, that is, (a) species richness, (b) the space (areas) for different types of landscapes that provide habitat for species richness, (c) the global water cycle; and (d) flow of the nutrient's nitrogen and phosphorus. These boundaries require different types of local, regional, national, and transnational cooperation in many different places to ensure the integrity of each unique area. These boundaries are set so that we

can keep an eye on the basis of life on Earth. There are no specific catastrophic trigger points in these boundaries, but rather a continuous deterioration in quality the further out the regional or local systems go.

(3) The third and final group deals with problems created during industrialisation and covers (a) the amount of new harmful pollutants, all of which are alien to the environment in which they occur, and (b) the amount of air pollution by aerosols, that is, small particles in the atmosphere that affect both air quality and weather systems. These systems are not natural but have arisen as a consequence of human activities over the last 150 years.

The framework for the planetary boundaries was updated in September 2023 to a Planetary boundary framework 3.0. What follows is a review of the nine planetary boundaries and an update on where we are now (see Figures 1 and 2).

(1) **Climate change**

Climate change is by far the most discussed topic in sustainability. So much so that some people mistakenly believe that sustainability is only about climate change. The phenomenon is called 'carbon tunnel vision' and the fear is that we will only manage to deal with this part of the sustainability crisis. This is understandable, climate change is spectacular in the scale and variety of its effects and 2023 has truly shown the shocking effects due to the El Ninõ effect overlaying the incremental effects.

In the field of monitoring climate change, the rise in the world's average temperature naturally takes centre stage. In 2015, the countries of the world came together to create the Paris Agreement under the leadership of executive secretary Christiane Figures. The agreement aims to limit the global temperature increase to below 2°C and to work towards the more specific target of 1.5°C. By the early 2020s, we have achieved 1.1–1.2°C, and the effects are evident in many parts of the world. Already in the early 2030s, we expect to cross the 1.5°C threshold. The reason for choosing 2°C is that, based on studies of Earth's history, this is the limit within which we should be able to keep our civilisation functioning. However, this depends on a number of trigger points that have recently been identified.

Trigger points are specific changes that have the potential to accelerate climate change irrespective of the CO_2 content. Several trigger points have been identified, but it is difficult to predict when these trigger points will be reached. The trigger points that are attracting the most attention are in the Amazon Forest, Antarctica, the Greenland ice sheet, the tundra, and global ocean currents. We are therefore in the frustrating position of not really knowing how far we can go. Hence, Antonio Guterres's metaphor of driving a car in the dark without lights through hairpin bends.

The driving factors for climate change are the emission of greenhouse gases including CO_2 and aerosols, and surface albedo (reflection of light). Before industrialisation, the concentration of CO_2 was 280 ppm. The planetary limit for 'safe operation' is set at 350 ppm. The CO_2 concentration in the early 2020s is between 410 and 420 ppm. Once we cross the

450-ppm threshold, we enter a high-risk zone. The 1.5°C limit is the basis for the establishment of a global carbon accounting system, which shows how much CO_2 we can still emit before we achieve the 1.5°C temperature threshold with a 50% probability. Attention is currently centred on milestones in our carbon budgets: when we reach peak CO_2 emissions, when we halve emissions, and when we finally stop emissions altogether. These budgets depend on a wide range of different assumptions. But it is worth noting that the time horizon is very short now. An idea of the order of magnitude is that if we are to have a 50% chance of staying below 1.5°C, world CO_2 emissions need to be halved by 2030 and eliminated by 2050.

From a project manager's perspective, this is a very short time. And still CO_2 emissions keep rising, so worldwide, we have not even reached peak emission.

(2) **Ozone layer depletion**
The depletion of ozone in the stratosphere received much media attention in the late 1980s. Ozone in the stratosphere (15–50 km out in the atmosphere) protects the Earth from UV light from the sun. Ozone is a volatile gas made up of three oxygen atoms stuck together instead of the usual two oxygen atoms. In the mid-1980s, an ozone hole was discovered over the polar regions. The ozone depletion was caused by the release of chlorofluorocarbons and halon gases (industrial gases used for cooling and other purposes). In 1987, the countries of the world met in Montreal and agreed on the Montreal Protocol, which banned over 100 different substances internationally. Since then, ozone concentrations in the stratosphere have increased again and the Montreal Protocol is now recognised as the most successful piece of international environmental legislation ever. Ozone depletion is now within its safe planetary boundary.

(3) **Ocean acidification**
A large part of the CO_2 released into the atmosphere is absorbed by the ocean. The chemical equilibrium that drives this transfer reduces the climate impact of our CO_2 emissions. Without the storage of CO_2 in the ocean, we would have crossed critical limits on climate change long ago. The problem is that CO_2 in water causes acidification of the ocean. Ocean acidification lies just below the boundary and is still rising. Among other things, acidity prevents or inhibits the formation of calcium carbonates, which are essential for corals, shellfish, and sponges. This means that marine life is threatened, slowly but steadily.

(4) **Loss of biodiversity**
Biodiversity is about species richness. The more different species there are on Earth, the better. All the different species fulfil different roles in different niches and thus form a robust interconnected ecosystem. Species have always appeared and disappeared. Through studies, it is possible to identify a kind of background rate at which species go extinct naturally. Under normal conditions, 0.1–1 species out of 1 million species become extinct per year. That corresponds to 0.00001–0.0001% loss of species per year. There have been five mass extinction events in Earth's history. The most famous is marked by the extinction of the dinosaurs 65

million years ago, when the Earth lost 50% of its biodiversity through a meteor hit. The planetary threshold for sustainable extinction rates has been estimated at 10 species per million per year – that is 10–100 times the natural level. The problem is that we are currently losing 100–1000 species per million species per year. One per 1000 per year means that one thousandth of all species becomes extinct per year. Again, we are far beyond what is sustainable, and there is an understanding among scientists that this is the sixth mass extinction event in Earth's history.

This parameter was in the recent update to the framework supplemented with measuring the parameter HANPP (human appropriation of net primary productivity). This parameter measures the fraction of the biological production on Earth that humans consume for food, paper, wood, fibre, and so on. The current HANPP parameter (30%) is greatly exceeding both the safe planetary boundary (<10%) and the upper end of zone of increasing risk (20%), hence we are already at high risk on this parameter.

(5) **Use of land**

The loss of area for nature around the world is happening at an enormous rate. Land is monetised in terms of urban, industrial, and agricultural growth. This means that many natural areas and thus habitats for animals, plants, insects, and so on are disappearing. This global boundary is also exceeded.

E.O. Wilson argues in his book 'Half Earth' that we should leave at least half of the Earth's surface covered by nature. The nine planetary boundaries framework estimates that at least 75% of the original forest area should be preserved. Unfortunately, we are already down to 62%. At the end of 2022, under very difficult negotiations, the UN (Kunming–Montreal global biodiversity framework) adopted targets at international level whereby 30% of land areas and 30% of marine areas must be under nature conservation by 2030. Today, the figures are 16 and 8%, respectively. The agreement is considered the 'Paris Agreement for nature'.

For the planetary boundaries for the use of land parameter is the remaining forest cover compared to potential forest cover for tropical, temperate, and boreal – which should be 85, 50, and 85%, respectively. The current estimates are (www.science.org, 2023) as follows:

- Tropical: Americas, 83.9%; Africa, 54.3%; Asia, 37.5% (critical limit 85%)
- Temperate: Americas, 51.2%; Europe, 34.2%; Asia, 37.9% (critical limit 50%)
- Boreal: Americas, 56.6%; Eurasia: 70.3% (critical limit 85%)

(6) **Fresh water**

Fresh water recently exceeded its limit. The definitions of the limits changed with the planetary boundary 3.0 framework and is now defined as streamflow as a proxy to represent blue water (surface and groundwater) – and root-zone soil moisture to represent green water (plant-available water). The parameter determines the fraction of land where the pre-industrial level is disturbed by humans compared to pre-industrial time. The limits are 10.2 and 11.1% respectively – however, currently the actual measurement is at 18.2 and 15.8%.

Many areas are already living under water stress and the number of areas is increasing year by year. Therefore, an increasing number of places will be at risk of experiencing severe water supply problems and will be depending on favourable weather conditions. Some years will for sure create serious problems. More extreme weather phenomena – both flooding and droughts – make water conditions more unpredictable and riskier. The more water stressed an area is, the less rare variation in weather – in statistical terms – will it take to bring the area into severe water crises. At least 50% of the world's population live under water-stressed conditions for at least 1 month of the year. This problem is projected to be worse in the coming years and decades.

(7) **Biogeochemistry**
Our over-consumption of nutrients used in food production ends up in large amounts in lakes, rivers, and oceans. Here, nutrients create polluted conditions where algal blooms lead to oxygen depletion and at its most severe remove the conditions for life in aquatic environments. For nitrogen, the amount lost to aquifers is almost three times higher than the planetary limit, and for phosphorus it is twice as high. Along with other types of water pollution, nutrients are destroying the environment for the life that should live in oceans, lakes, rivers, and streams. Despite all the work on wastewater treatment plants this is clearly still not sustainable.

(8) **New substances**
Emerging substances consist of thousands of man-made substances that are harmful to the environment: microplastics, pesticides, pharmaceutical residues, personal hygiene products, and all sorts of industrial chemicals. It is estimated that there are 350 000 man-made chemical substances in the environment. These substances pose a huge problem. There have long been uncertainties regarding the definition and location of this limit; however, in 2022, scientists definitively determined that this planetary boundary has been significantly exceeded.

(9) **Atmospheric aerosols**
The last global limit relates to aerosols. Aerosols are small particles which, due to their small size and weight, can remain suspended in the atmosphere. This limit has long been difficult to quantify. However, with the latest framework a limit has been set based on interhemispheric difference in AOD. AOD is the aerosol optical depth, that is, how much of sun light is reduced by absorption and scattering due to aerosols. The current estimate of the parameter is at 76% of the limit – and hence this boundary has still not been exceeded. However, this limit still needs further research to be more precise – especially the geographic distribution and the effect of this is not yet well understood.

There are at least two negative effects of aerosols; one is that they enter the airways of humans and animals, where they cause a number of respiratory diseases, many of which can be fatal. In addition, aerosols have various effects on the atmosphere. In some contexts, they are believed to have a cooling effect and can thus counteract global warming.

So, where do all these global effects leave us? This is illustrated in Figure 1:

(A) Six boundaries have already been exceeded: biodiversity, climate change, chemical pollution, biogeochemistry, fresh water, and land use.
(B) Three effects are within their boundaries: ocean acidification, aerosol loading, and stratospheric ozone depletion.

It is this overview that constitutes the fundamental planetary issues that are at the root of the most severe threats of our environmental sustainability crises. And it is our understanding of these limits that makes it possible – potentially – to steer back to a 'safe space'. However, as illustrated in Figure 2 the situation is worsening over time from 2009 to 2023.

We don't know if we will succeed in resolving these crises and creating a good state for the Earth let alone how many problems we must go through before. Maybe by 2030 we will know if we succeeded in pushing some of the parameters towards a safer place. Currently, the focus is on whether we can successfully reduce CO_2 emissions by half and whether any trigger points are crossed before that. But we need to increase our attention to the current loss of biodiversity which is becoming increasingly critical. This is the time to really do whatever we can to stay within the nine boundaries.

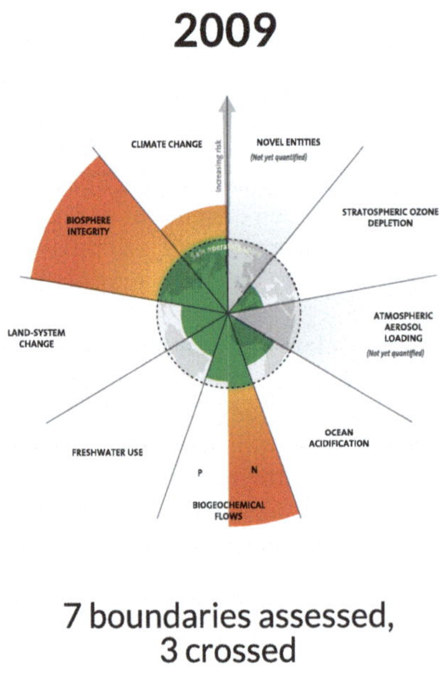

Figure 2 Development over time for the nine planetary 2009–2023 (*credit*: Azote for Stockholm Resilience Centre, Stockholm University. Based on Richardson *et al.* (2023), Steffen *et al.* (2015), and Rockström *et al.* (2009).

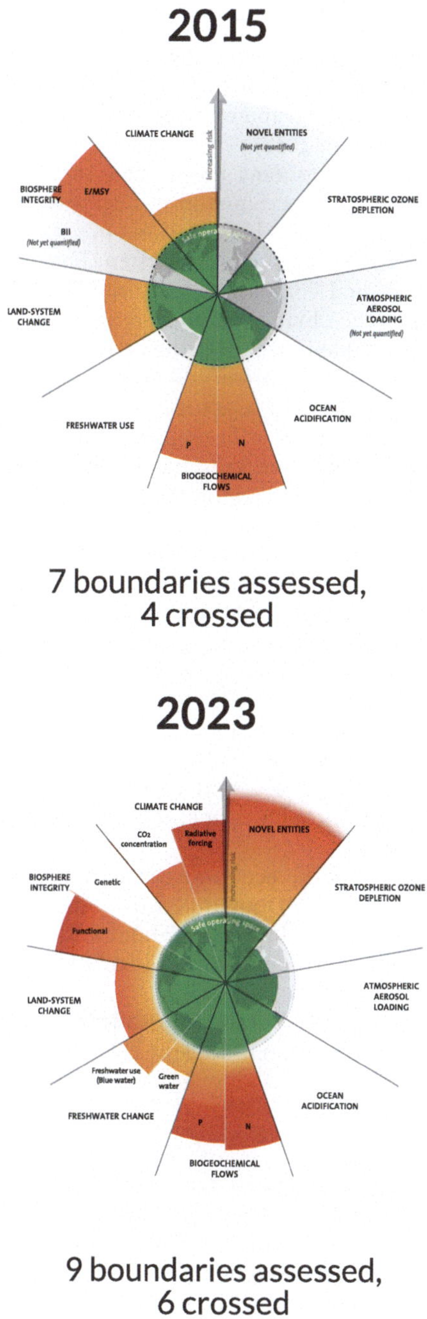

Figure 2 *(Continued)*

DOUGHNUT ECONOMICS

The way we have organised the economy is one of the main barriers to a sustainable future – some name this hyper-capitalism. Economist Kate Raworth has proposed a new theoretical framework for understanding and changing our economic thinking (Raworth, 2018).

The organisation of the economy has a huge impact on the feasibility of sustainable measures in projects. Short-term business case calculations often yield the impression that sustainable solutions are too expensive. Therefore, less sustainable solutions are chosen far too often. The problem is that they are too expensive for individual companies and individuals, but from an overall societal perspective it is the obvious – and, overall cheaper – solution. It is our fragmented and myopic behaviour that is the expensive solution.

The economy as it is currently organised creates enormous inequality in the world. This makes it more difficult to achieve solidarity, which is a prerequisite for the difficult transition to a different sustainable society. Politics and economics are closely linked. In order to have different politics, it is necessary to think and work differently with economics. One possibility is to start working with different economic principles.

Even if – from a project point of view – we cannot change the major global economic systems, there are still opportunities within the current economic system to change the way we act if we think more creatively about, for example, negotiation economics. Here it is a basic psychological observation that we are more concerned about not losing a small amount of money than we are about gaining a larger amount. This often gets in the way of how we negotiate, and the consequence is that a lot of money is left on the table in negotiations between negotiating organisations. Even with the current economic mindset, we currently don't manage to see and realise all the 'win–win' potentials that exist, potentials that, through cooperation, could be used to infuse more sustainability into projects.

Let's try to have a look at the economy through the eyes of economist Kate Raworth. She starts her book 'Doughnut Economics: Seven Ways to Think Like a 21st Century Economist' by talking about a student revolt in economics programmes at universities around the world in 2014, following the economic crisis.

Kate Raworth quotes the International Student Initiative for Pluralism in Economics (2014) 'An international student call for pluralism in economics' (Raworth, 2018):

In an open letter, the students wrote:

> The teaching of economics is in crisis too, and this crisis has consequences far beyond the university walls. What is taught shapes the minds of the next generation of policymakers, and therefore shapes the societies we live in ... We are dissatisfied with the dramatic narrowing of the curriculum that has taken place over the last couple of decades ... It limits our ability to contend with the multidimensional challenges of the 21st century – from financial stability to food security and climate change.

A fundamental problem with current economic thinking is that it focuses as its goal the maximisation of each country's gross domestic product (GDP). Looking historically at the origins of economic thinking, it started as far back as Ancient Greece, when Xenophon posed the question: how should a household best manage its resources? The scope of the question was subsequently expanded to include how Athens should best manage its resources. Later, Adam Smith (Scottish economist and moral philosopher) further expanded the scope of the question to include an entire nation when he wrote 'The Wealth of Nations' (Smith, 1776), a book that has since become a kind of primer for establishing the fundamental principles of economics.

From Adam Smith, the best-known idea is that the market is guided by an invisible hand that ensures that the overall economic system grows as much as possible. This idea was complemented by the idea of trickle down economics, or, as Adam Smith called it, 'a rising tide raises all boats'. The idea was that as a community became richer, everyone would automatically become richer. This has led – despite much criticism – to a very narrow focus on the GDP of individual countries and thus an understanding that all countries are in principle in competition with each other to achieve the highest GDP. In today's world, this idea has become the engine that is driving the world closer and closer to our sustainability problems.

Kate Raworth sets out to describe a different way of thinking about economics, one that was not on seeing countries in competition with each other, but instead trying to see what it is that we fundamentally want economics to do.

During the 1990s, the UN worked on the development of an alternative to GDP, the so-called human development index (HDI). The HDI includes aspects such as long and healthy lives, educational standards, and a decent standard of living. This index found that the 'winning countries' were those that had found democratic ways to convert their country's wealth into progress and well-being for all. The Scandinavian countries in particular came out as the winners in HDI measurements. The problem with the HDI, however, was that these countries often had huge ecological footprints. Their populations were thriving, but nature and the foundations for future life were under pressure both at home and through the pressures they were creating abroad through their own consumption. It was clear that there was a need to complement HDI with an understanding of environmental constraints. Kate Raworth came up with the idea of combining the ideas of a good and dignified human life with those of planetary boundaries. She illustrated this by using a doughnut as a measure of good economic principles.

The basic idea of the doughnut economy is that a society that is sustainable meets two sets of boundaries illustrated by the inner edge and the outer edge of the doughnut (see Figure 3).

In the inner part of the doughnut, no human being should be left 'in the hole'. The gap is defined by parameters that constitute the minimum social foundations for a good healthy human life, as defined by the UN SDGs: adequate access to water, food, health, education, income and labour, housing, networks, energy, peace and justice, political voice, social equity, and equality. No one in our society should fall short on these social parameters. If these parameters

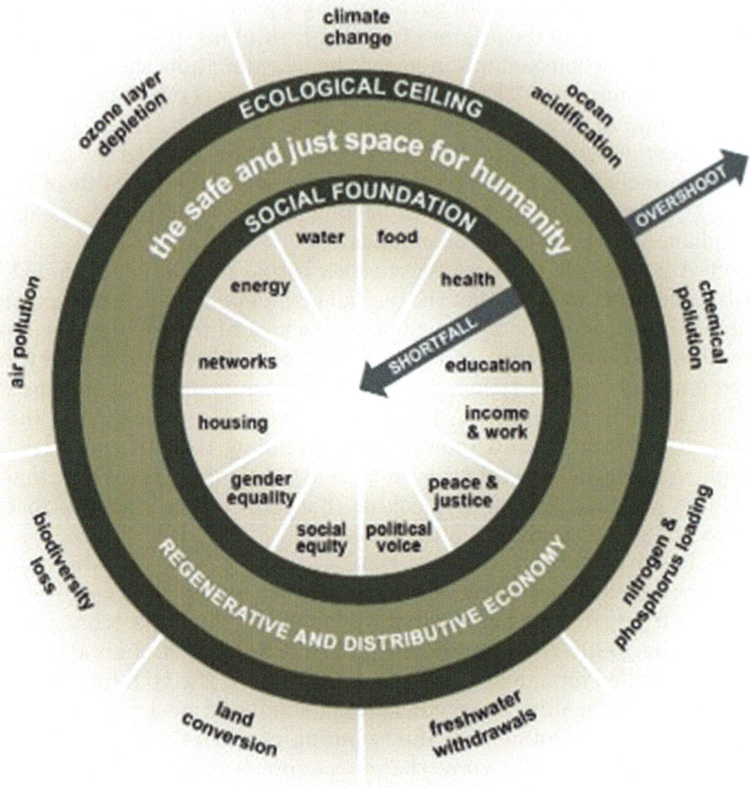

Figure 3 Illustration of the doughnut economy. As a global society, we must avoid both ecological overshooting of the nine planetary boundaries, and socially avoid falling below the limits of good social foundations for humans (*source*: figure reproduced from Raworth's book and work).

are not met, people will suffer, and unrest will follow – causing environmental problems that threaten the planet as well as local areas. Because the world is so globally interconnected, it is not possible, as a few decades ago, to keep unrest in one area. Refugee flows, terrorism, and wars destabilise the whole world – both socially and increasingly obviously also environmentally. Over the past few decades, the ties between nations, industries, organisations, and all people have become very visible to all – and our travels to distant countries have helped everybody understand that even people far away must be treated with respect.

The challenge is that this transition to a global society that fulfils the needs of all must take place within the ecological ceiling that ensures that the planet can continue to act as a base for human life and our societies. The idea of the nine planetary boundaries is used as the outer edge of the doughnut: we must stay within the limits of climate change, ocean acidification, chemical pollution, nitrogen and phosphorus loading, clean water, land use, biodiversity, the ozone layer, and aerosol air pollution. All this must happen at the same time as the world's population grows, and corona epidemics, wars, rebellions, and so on play out.

The method works as a simple way to understand very complex interrelationships. The approach has been tested in several cities. The most ambitious trial is under way in Amsterdam. The results yield a good indication of how to work with the new economic mindset in practice (https://doughnuteconomics.org/Creating-City-Portraits-Methodology.pdf, 2023).

Key questions asked to citizens in Amsterdam during the trial:

- How can your city become a home for people who thrive while ensuring the well-being of all and the health of the planet?
- What does it mean for city citizens to thrive?
- What would it mean for the city to respect the well-being of other people around the world?
- What does it mean for this city to thrive in its natural habitat?
- What does it mean for the city to respect the health of the whole planet?

In Amsterdam, they found a nine-step method to answer those questions and turn the answers into actions. The steps were:

(1) Mirror – what does it look like today.
(2) Mission – create a powerful vision from the mirror for a system that thrives better.
(3) Mobilise – gather change agents and give them an opportunity to engage.
(4) Map – provide an overview of initiatives and strategies.
(5) Mindset – apply the new doughnut mindset.
(6) Methods – use other methods that have been proven to work already.
(7) Momentum – creating momentum through iterative and reflective processes.
(8) Monitor – monitor progress and communicate it.
(9) Mmmm! – makes it fun and creative to be part of.

The local action plans were relatively simple and included activities such as water reuse and hiring more qualified teachers in schools. The global issues were more difficult to translate into action, but ways were found anyway. The nine planetary boundaries were translated into the city's fair share of the footprint. This was not trivial but still feasible to the extent that targets can be set for the near future, such as 'Reduce the city's CO_2 emissions to 55% below 1990-levels by 2030 and to 95% below in 2050' or by looking at a snapshot such as 'The Netherlands has the highest water footprint in Europe, with almost 90% of total water use embedded in imports such as meat, cotton and food'.

When it comes to global social responsibility, there are many interactions that make it difficult to set clear targets. Instead, the city tried to mirror the effects of choices in the city with quotes that highlighted the impact of the choices of the city and its inhabitants. In relation to electronics, this quote was chosen: 'We all have problems with our lungs and pain all over our bodies' – female cobalt miner from the Democratic Republic of Congo. Another quote related to the consumption of clothing was 'In Asia, women garment workers often face forced overtime, sexual harassment and being fired if they are pregnant'. The quotes were intended to provide a concrete understanding of the consequences of the consumption choices of the city's citizens. (This description is reproduced from the experiences reported at https://doughnuteconomics.org/Creating-City-Portraits-Methodology.pdf, 2023.)

Doughnut economics has a huge momentum right now in the development of economic theory. The theory is making its way into mainstream economics textbooks, which explain the evolution of economic thought from Adam Smith to Kate Raworth.

The aim of Raworth's doughnut is to create a society where there is a social foundation – a minimum level below which no one falls, and a global ecological ceiling above the maximum pressure we can put on natural resources – the planetary boundary. Between the two limits lies a safe and just society. In this way, a metric has been created to steer the economic system in a way that is both environmentally and socially sustainable. A society that works is now defined.

THE SUSTAINABLE DEVELOPMENT GOALS

In his book 'The SDGs – The most important topic at the agenda', Professor of Management Steen Hildebrandt states that '*You cannot lead without goals, and you cannot have goals without leadership*'. He also stated that the SDGs are the greatest opportunity and leadership challenge that has ever existed (Hildebrandt, 2018).

The SDGs are the global community's response to the ecological and social problems facing the world. They are a template for achieving 'a better and more sustainable future for all', with clearly defined goals and a set deadline of 2030. The goals are summarised in a resolution called Agenda 2030, which was adopted in 2015 by the UN General Assembly. They are to be seen as the next step after the Millennium Goals but differ significantly in their approach. The (old) UN Millennium Goals were aimed at what were then referred to as developing countries. With the SDGs, there is a recognition that all countries are developing countries, and that the development path ahead is more difficult than previously realised.

The goals that were adopted (see Figure 4) are broad comprehensive goals: no poverty (1), zero hunger (2), good health and well-being (3), quality education (4), gender equality (5), clean water and sanitation (6), affordable and clean energy (7), decent work and economic growth (8), industry, innovation, and infrastructure (9), reducing inequality (10), sustainable cities and communities (11), responsible consumption and production (12), climate action (13), life below water (14), life on land (15), peace, justice, and strong institutions (16), and partnerships for the goals (17).

Two years later (2017), the broad goals were made more tangible through actual targets and indicators to measure the progress on each SDG. For each

Figure 4 UN Sustainable Development Goals.

SDG, there are 8–12 targets, each with one–four indicators. Most targets have an end date between 2020 and 2030. The goals are monitored every 5 years, that is, 2020, 2025, and 2030. Some of the targets are well defined, whereas for others it has been necessary to use targets for which a measurement methodology was not yet available.

Through the SDGs, the nations of the world have succeeded in creating a vision for a better and more sustainable world in the future. This is backed up by clear priorities and targets for how fast we need to move forward at the global level on a wide range of parameters. The targets have been made concrete and measurable and are monitored by statisticians so that it is possible to intervene if there is insufficient progress on each parameter – intentions are not enough.

On the one hand, the concrete targets are an impressive diplomatic and procedural achievement, the result of many negotiations. On the other, the goals are not always so unambiguously helpful when you try to apply the goals as a country, an industry, or a company or right down to project or individual level. In practice, it's evident that few individuals work directly on the parameters of the SDGs. Instead, it's crucial to consider the numerous preconditions for these goals, as those are the elements you'll actually be engaging with. This indirect work is necessary for change to happen. In this way, the SDGs serve as a training ground for everyone in holistic thinking and action. The SDGs represent a renewed mindset about management and prioritisation.

We therefore need to practice two things with the SDGs. Firstly, is training in thinking systemically, and based on this idea, it makes sense that the SDGs are most often presented as a collection of all the goals together. Through this lens, it is not about the individual goals, but about the overall vision for a 'better and more sustainable world'. It is about the two basic principles of the vision, which are 'do no harm' and 'leave no one behind'. This is a way of acting that we are not used to. It requires training. Secondly, at the same time, we need to stick to an approach that is about being concrete and ensuring that we are actually moving towards the goals. There needs to be measurements. Therefore, we both need to understand the big picture, but also need to pay attention to the details of each goal.

It is a huge shift in our mode of decision making to take global responsibility in all of the billions of decisions made everywhere every day. Most people probably experience this sense of 'everything is connected' from time to time. For example, when we realise that a problem with a new virus in China is suddenly the problem of the whole world. Then we can feel how we are all breathing the same air. But as humans, we are often quick to forget this again in the face of everyday problems that calls for quick local solutions. An often-recurring argument – said or unsaid – is this: 'what good are my efforts, I can't save the whole world?'. The difficulty is that it can be hard to distinguish our own individual positive or negative contribution in this global system of 7.5 billion people. Nevertheless, it is the sum of each individual's positive or negative efforts that will determine whether and how our transition manoeuvre will work. No one can do it alone.

SUSTAINABILITY AND THE WATER SECTOR

When zooming in on the sustainability of constructing water infrastructure, be it structures for water extraction, water treatment, water distribution, sewer systems, or wastewater treatment, it is helpful to distinguish between two levels of sustainability: (1) system sustainability and (2) design and construction sustainability.

(1) System sustainability is primarily handled during or before the early planning phase of projects and have a systems design process at its centre. The aim is typically to extend, replace, or redesign a city's existing water infrastructure. In the planning stage important principles of systems design are applied with a strong focus on the context of the technical system as well as the environment and all the stakeholders including customers, users, authorities, and non-governmental organisations (NGOs).

(2) Design and construction sustainability on the contrary is handled in the later planning phase and the execution stage, where the actual infrastructure in terms of pipes, plants, pumps, buildings, and so on are contracted and built.

The required types of knowledge are different in the two types of sustainability. Water system sustainability (1) relies heavily on the 'water sector knowledge domain', whereas design and construction (2), that is, the actual execution of the planned projects relies primarily on 'the construction knowledge domain'. Generally, the water domain knowledge is used in the planning phase of projects, whereas construction domain knowledge is more relevant when the projects have been defined and is to be executed. Therefore, it is important to have people based in both fields in planning and projects departments in utilities and in consulting companies.

The purpose of the planning departments in utilities worldwide is to take a systems design view and thus have a thorough understanding of the current water infrastructure and find out how to extend, upgrade, or refurbish the current system while at the same time thinking ahead to ensure that the new additions prepare the system for the future. This is why innovation takes such a large (but also often problematic) space in the water sector. There is a need to create the new thing that is better than the old. However, at the same time it needs to work in the already established 'old system'. The lifetime of new assets is often somewhere between 20 and 100 years. And in organisations where stability of supply is the primary concern, the risk appetite is limited as to applying new principles and technologies. On the contrary, renewal is required for the system to develop. Therefore, it is all about finding a good balance.

In utilities the following circular process is playing out repeatedly throughout changing parts of the system:

(1) Needs assessment
(2) Feasibility study
(3) Conceptual design and planning
(4) Permitting and approval
(5) Detailed design and engineering
(6) Construction
(7) Commissioning
(8) Operation and maintenance
(9) Monitoring and evaluation
(10) Decommissioning or upgrading

Looking at time constants in a water system, the work taking place in operations often requires planning a few days or weeks ahead, maintenance requires monthly or annual scheduling, while the carrying out of planning and projects have years or even decades as the main time horizon.

In the circular process above, the planning phase includes steps 1 through 4, whereas the project stage typically involves stages 5 through 7 plus step 9. Finally, step 10 at some time merges into step 1 and the circular process starts over.

Before looking into different tools for sustainability, it may be relevant to consider another distinction to be made when working with sustainability. The question that soon gets both relevant and difficult is: on a scale from not very sustainable to very sustainable, what should this particular project aim for?

As illustrated in the discussion above on SDGs and doughnut economy, this question is far from trivial. Besides from all the parameters, that need to be considered and balanced, there is still the question of where to set the benchmark for sustainability. A distinction that becomes increasingly clear these years (in the first half of the 2020s) is the distinction between three domains of aims: degeneration, sustainable, and regeneration (see Table 1).

Although it may be difficult to create projects that are entirely in one of these three categories on all parameters, it may still be helpful to discuss where the general aim of the project is. This will help the team become clear on methods and technologies.

Sustainable project management is relevant both in the planning and the execution stages. A key task for a sustainable project manager is to ask questions in both phases and ensure follow-up on sustainability. However, as the approaches at the two levels are so different, it is relevant to look towards

Table 1 Three domains of aims: degeneration, sustainable, and regeneration.

Degeneration	Sustainability	Regeneration
Net negative	Neutral	Net positive
Shareholder centric	Techno-centric	Stakeholder centric
Short term	Medium to long term	Long term
Growth	Growth	Improvement of life
Competitive		Collaborative
Laws and regulations	SDG or doughnut economy	SDG or doughnut economy
Organisation as machine		Organisation as living organism

different types of tools to manage the sustainability impact of the project and possibly also consult with people with different theoretical backgrounds.

Water stewardship

The planning phase of the project or the planning department is charged with the purpose of year by year transforming the urban water system towards greater degrees of sustainability in terms of using, for example:

- Retrieving water in a way that is better for the ecosystems
- Using less energy in water treatment
- Loosing less water from the water distribution system
- Reducing combined sewer overflow
- Reusing wastewater
- Treating wastewater better for lower effluent of nitrogen, phosphorous, organic matter, pharmaceuticals, and so on
- Reusing sludge for energy production
- Applying blue-green solutions, and so on

Sustainability becomes practical through projects. The process is rarely as straight forward as one would imagine or wish. There are a lot of different parameters to consider and balance to stay clear of planetary boundaries and the social ceiling.

Johan Röckström, who has been a primary driver in the development of the planetary boundaries, is increasingly speaking of stewardship – either as Earth stewardship or stewardship of the planetary boundaries. For the water sector the obvious next step is to think, work, and act in terms of water stewardship. The word 'water stewardship' has been known for several decades but has not yet lived up to its potential as a new way of understanding a role of local caretaker for water, a role that local municipalities or water utilities could play and thereby improving water availability and quality for both people and wildlife (see e.g. Ingildsen, 2020).

A central framework for water stewardship is the idea of integrated water resource management (IWRM).

A generally applied definition of IWRM is:

> Integrated Water Resources Management (IWRM) is a process which promotes the coordinated development and management of water, land and related resources in order to maximise economic and social welfare in an equitable manner without compromising the sustainability of vital ecosystems and the environment. (www.unep.org, 2023)

IWRM has been accepted to the extent that it has been included as one of the indicators under SDG 6. The SDG text is

> Target 6.5 aims to implement integrated water resources management (IWRM) at all levels. Indicator 6.5.1 tracks the degree of integrated water resources management (IWRM) implementation, by assessing the four key components of IWRM:
>
> - Enabling environment.
> - Institutions and participation.
> - Management instruments.
> - Financing.
>
> It takes into account the various users and uses of water, with the aim of promoting positive social, economic, and environmental impacts at all levels, including the transboundary level, where appropriate. (www.unwater.org, 2023)

Another important tool for keeping track of a company's water stewardship efforts is the water stewardship certification under The Alliance for Water Stewardship.

The Alliance of Water stewardship offers a complementary approach to IWRM. The method takes on a bottom-up approach, which means it can be applied by singular actors – and may therefore be a good place to start for, for example, utilities. However, so far it has still only been applied by major water consuming companies. The Alliance of Water Stewardship is a global membership organisation including businesses, NGOs, and the public sector. The organisation is starting a movement where members through the adoption of the certification system 'the International Water Stewardship Standard' increase their awareness of their own water issues combined with an understanding of the system they are part of. The approach embraces the 'learning-by-doing' principle by including a systematic approach to continuous improvement actions, which is a prerequisite when working with water in an integrative way.

A fundamental principle is the belief that if major water users gain a sufficient understanding of their own water use and impacts water in nature and commit to publish these results transparently, then they will find the will and ways to improve. The public element and some of the criteria for the certificates ensures facilitation of coordination with other water users within the wider water catchment area.

'The water stewardship standard' covers the areas of good water governance, sustainable water balance, good water quality, important water-related areas, and safe water, sanitation, and hygiene (WASH) for all.

Good water governance means to have a governance system implemented internally so that responsibility is clarified and a framework for action is implemented. This framework for actions must include a system for maintaining and updating water sustainability-related information.

Sustainable water balance means that the user's direct water consumption is withdrawn from natural resources in a way to ensure that the water resources are continually replenished. Similarly, the consumers need to investigate the sustainability of its significant indirect water use when possible.

Good water quality status means understanding and taking co-responsibility for the water quality of the aquifer from where the water is abstracted.

Important water-related areas mean identifying impacted areas with special water interests, either for water abstraction or wastewater effluents. In these areas, the water steward must comply to effluent standards as well as engage in the quality of the water-related area as a prerequisite for thriving eco-systems.

WASH is shorthand for safe water, sanitation, and hygiene for all and represents the social responsibility for other less powerful water users. The criteria are related to the concept of 'water as a basic human right'.

The two frameworks, IWRM and water stewardship certification, complement each other effectively in providing a comprehensive perspective on water bodies within specific local areas. Despite this, it is evident that a significant journey remains before we achieve a highly diligent level of water stewardship. The gaps are extensive and encompass a general comprehension of water's role in local nature, as well as its significance for stakeholders. Both of these knowledge domains require reinforcement through dialogue, measurements, and models. Most utilities and major water consumers tend to focus on their own needs rather than adopting a holistic understanding of water and its local usages, which is essential for true water stewardship.

The development of digital twins will forge new understanding through models of groundwater, surface water, wastewater treatment plants, sewer system, and tap water distribution systems. At some point we will be able to link it all together and hence have a much broader and deeper understanding of the full system and all its interactions. This will enable us to design better systems that are aligned with all the stakeholders including those in nature. The models can also be used to ensure that the water system designs are tested thoroughly under all kinds of conditions such as heavy rainfall and droughts – as well as predicted future situations including the inevitable effects of climate change. The digital revolution in the water sector promises a lot of possibilities for proficient water stewardship.

Feasibility studies is another area that needs to be further developed, so that it is not only economic feasibility that is mapped, but also environmental and social feasibility. A simple tool and hence a good starting point is provided by the SDGs – where the progress or setbacks on each SDG should be described and if possible quantified. Making a kind of heatmap of a project on the SDGs can be used as a supplement for the traditional economic evaluation.

An interesting development that is seen in several countries is the idea of giving major water bodies legal personhood status. In countries as diverse as India, the USA, Brazil, Peru, and Canada, Ecuador, Bangladesh, Bolivia,

Uganda, Australia, and Netherlands various kinds of experiments of giving voice to rivers and other water bodies – or even specific animal species – are being carried out. It is important to ensure that the entire water system reflects what is best from the perspective of water itself. It – at least to some extent – provides a kind of promise to salvage the often-fragmented body of legal laws meant to protect the environment. A complementing trend that seems to emerge is research, understanding, and implementation of indigenous water management principles and theories.

Regardless of how you approach the planning process, it is important to gain deep systems understanding and supplementing the traditional engineering practices with books on systems design. Applying systems thinking can be extremely beneficial for generating innovative ideas to address various situations within utility or water catchment areas.

Design and construction

In the construction industry there are a number of well-documented frameworks for sustainable buildings. The ones, that are most widely used include DGNB (German standard), BREEAM (UK standard), LEED (US standard), HQE (French standard), Level(s) (new EU standard), Green Star (Australian standard), EDGE (IFC standard (World bank)), Estidama Pearl Rating System (United Arab standard), CASBEE (Japanese standard), The Three Star standard (Chinese standard, actual name: The Ministry of Construction's Green Building Evaluation Standard), and the Green Building Code (Indian standard). As we can see, various countries have developed different standards for how to ensure sustainability in construction. It will obviously be going too far to describe all of these building certification systems in this book. Instead, you need to find the one that is the accepted standard in your country and study that thoroughly.

To give an idea of what such systems may entail, the following is a brief description of the German DGNB standard. The DGNB standard supports a systematic approach to construction from start to finish in projects. Different levels of certification are possible: silver, gold, and platinum. The DGNB system for new buildings and major renovations is based on a point system, where increasing amounts of points are required to achieve respectively bronze, silver, gold, and platinum.

It is possible to score points within the main topics: environmental quality, economic quality, social quality, and the cross-cutting technical quality, process quality, and area quality (see Figure 5).

When mapping the relationship between DGNB and the SDGs, we see that construction primarily affects the following SDGs: 3: Good health and well-being, 8: Decent work and economic growth, 11: Sustainable cities and communities, 12: Responsible consumption and production, 13: Climate action, and also, to a somewhat lesser extent, 5: Gender equality, 6: Clean water and sanitation, 7: Affordable and clean energy, 9: Industry, innovation, and infrastructure, 14: Life below water, 15: Life on land, 16: Peace and justice strong institutions, and 17: Partnerships to achieve the goal (see also Figure 6).

Figure 5 DGNB is based on six domains: environmental quality, economic quality, and social quality, as well as the cross-cutting domains: technical quality, process quality, and neighbourhood quality.

Figure 6 DGNB is based on 15 principles for circular building projects.

Points are awarded based on 36 criteria. A large proportion of the points available under DGNB include work that – in the context of this book – can easily be thought of as 'good project management'. It makes sense that the process (process quality) of thinking and acting wisely in the planning and execution phases is given high value (12.5% of possible points), as it is often possible to achieve a significant improvement in sustainability simply by thinking and acting reflectively.

These include, for example, tasks carried out during the planning phase:

- Strategic choices and assessment of project character
- User communication
- Quality of project preparation
- Sustainability in public procurement
- Guidance on maintenance and operation of the building
- Architectural quality process

As well as tasks carried out during the execution phase:

- Construction site and construction process
- Documentation of quality of execution
- Commissioning
- User communication

However, the tasks also include life-cycle assessment (LCA), total cost of ownership, flexibility, and adaptability, dismantling and recycling, and much more. It is not necessary to aim for DGNB certification as such. Alternatively, you can choose to be inspired by the DGNB certification process and try to fulfil the requirements without going through the actual certification process.

PRACTICAL SUSTAINABILITY TOOLS

The creation of the SDGs is an impressive story of an achievement the like of which humanity has never before attempted or succeeded. At the same time, it is a teeth-grindingly frustrating story of global change that seems far too slow compared to the needs of the planet. On the one hand, it is a story of hope for a better future, and on the other, success can seem improbable, threatening the future of people and nature. There is truth in both the optimistic and worried approaches.

You can – easily – think: 'what is the point of me putting effort into this project? My project has just a grain of sand of significance on one scale or another'. But that is precisely the point of the SDGs. The sustainability problem is solved by millions of grains of sand on the right side of the scale. There is no single approach that solves the entire problem. The problems have been created by billions of daily actions all over the Earth, and the problems if solved will be solved by billions of daily actions all over the Earth. As a project manager, adopting a very pragmatic approach is crucial: what actions can I take?

It is wise to remember the words of Václav Havel (Czech writer, playwright, and politician): '*Hope is not the conviction that something will turn out well but the certainty that something makes sense, regardless of how it turns out*'.

The SDGs are shared objectives that we collectively strive to achieve at every level. The goals are questions posed to each and every one of us: What can you do? How can you contribute? All individuals can contribute through their private choices of food, mode of transport, fridge, donations, rubbish disposal, and so on – and all the other choices we make every day.

For a project manager, there is a need for a greater understanding of how a project affects sustainability – how does the way I treat water locally affect sustainability locally and globally? Sustainability is a specific science. There are many underlying principles that need to be understood and these principles translate into lots of different options in the project process. Sustainability is embedded in everything we do throughout our lives and this sense of 'everything is connected' is different to navigate than most traditional engineering challenges. Recognising it requires a change in our perspective.

Some additional tools and concepts that can be helpful in this journey are explained below.

Life-cycle analyses

LCA is a methodology used to assess and compare the environmental impacts of a product or service throughout its life cycle, that is, it includes raw material extraction, production, use, disposal, and recycling. The basic principles of LCA are:

(1) LCA includes a system boundary that defines the scope of the analysis and determines which activities and impacts are included in the assessment. LCA evaluates the environmental impacts of a product throughout its life cycle, including effects such as greenhouse gas emissions, air pollution, water pollution, and land use.

(2) A functional unit of a product is a reference quantity used to compare the environmental impacts of different products. It is used to normalise the results so that the impacts of different products can be compared on a common basis.

(3) Accuracy and reliability depend on the quality of the data used in the assessment. LCA practitioners need to carefully evaluate the data sources used and ensure that the data are relevant, accurate, and up to date. In addition, it is important that the analysis is thorough and consistent so that the results can be understood and compared by different stakeholders. In addition, assumptions, methods, and data sources must be clearly documented and declared so that they can be checked and recalculated.

Circular economy

The idea of circular economy is to close the loop for all products so that nothing ends up as waste, that is, everything is reused. In practice, the circular economy works on the basis that all the ingredients of a product can be reused. This ultimately means that there is no point of consumption, everything returns to use again and again. Circular economy is similar to the concept of cradle-to-cradle, where the end of a product lifetime is the beginning of the next cycle of the product lifetime. This is in contrast to the industrial mindset of cradle-to-grave products, which have a traditional linear lifecycle where products eventually end up as waste.

Biomimicry

Established as a concept by researcher Janine Benyus (2002), biomimicry is a practice that learns from nature's design to solve human design. The idea is that we can learn from the principles that nature has developed over billions of years of evolutionary design. Examples from the construction industry are self-repairing concrete using bacteria (e.g., bioconcrete), solar panels that mimic photosynthesis (e.g., SunFlora), energy-efficient buildings based on ideas from termite nests using passive cooling (e.g., Termitemaster), and drought-resistant landscaping (e.g., Dryscape or Biogarden). Biomimicry is an innovative design method that uses nature as a model and mentor for products and services. The design methodology has sustainability and holistic thinking at its centre – always starting from the question: how would nature solve this problem?

ISO 14001

ISO 14001, like the other ISO standards, is aimed at organisations and provides the requirements for a business management system that ensures that the environment is addressed in a practical way throughout all business processes. Businesses that are or are working with the ISO 14001 standard will usually have specific requirements for project implementation in their organisation.

B Corp

B Corp is a certification scheme managed by a global NGO. Through this scheme, companies can be certified and become B Corp companies. B Corp is becoming more and more common in Denmark. Worldwide, more than 3500 companies are certified in all kinds of industries.

Environmental impact assessment

Environmental impact assessment (EIA) is a method for mapping all the environmental consequences a given project may have. EIA is a process used to systematically evaluate the potential environmental effects of a proposed project, development, or activity. The goal of EIA is to identify and assess the potential positive and negative environmental impacts of a project before it is undertaken. This helps decision makers, stakeholders, and the public to understand the potential consequences of a project on the environment and make informed decisions. Such analyses are widely used in regulatory processes as a basis for various permits.

MAIN POINTS

The concept of sustainability often sparks debate, yet many individuals lack a comprehensive understanding of its true essence. As soon as we try to make projects sustainable, it becomes clear that it is not as simple as we imagined.

For many, there is a perception that it is simply about reducing CO_2 emissions, but it is far more complex than that.

Successful sustainability requires the ability to think holistically, that is, with an understanding of all the factors that influence decisions, and long-term. At the same time, this 'holistic thinking' must not leave us paralysed by complexity, rather it must be translated into concrete action with actual measurable results. It is necessary to invest time and effort in reading about and understanding sustainability to understand what is at the centre of sustainability crises in order to act wisely on projects. It is easy to believe that everything with a sustainability label attached is sustainable, but it depends on the context and alternatives. Similarly, it's also easy to become cynical and think it's all greenwashing – that's just the other side of the fence and not helpful either. There is a need to take sustainability seriously as a multi-disciplinary discipline and do what you can from there.

There are nine planetary boundaries: climate change, biodiversity loss, biogeochemistry, ocean acidification, land use, clean water for drinking, ozone loss, atmospheric aerosols, and chemical pollution. Most of these limits have either been exceeded or are being exceeded at a rapid pace and with drastic consequences for life on Earth. Ozone is one notable exception, where we have seen that it is indeed possible to reverse the trend through international cooperation.

The doughnut economy is a different way of thinking about the economy. It is a change from the old GDP-optimising economic paradigms. The doughnut theory got its name because it resembles a doughnut, where the outer border is the nine planetary boundaries, whereas the inner ring is a series of social minimum limits that must apply for all people to live a dignified life. This way of thinking about economics has become highly relevant as a measure for how the world can become sustainable and operated in a sustainable fashion.

The conditions described in the planetary boundaries and the doughnut economy are translated into the SDGs which is a management tool for the whole world. The SDGs consist of 17 goals with a set of targets and indicators and was adopted in 2017. It is a hugely ambitious project, consisting of an overall vision of 'do no harm' and 'leave no one behind' and have clear targets that cover the whole planet. Every 5 years, there will be a review to see if each country – and therefore the world as a whole – is moving in the right direction and fast enough to achieve the goals, most of which have a deadline of 2030.

Water stewardship and the DGNB scheme are further concretisation that may be useful for the water sector. Water stewardship is especially useful as a tool for achieving sustainability in the planning processes of utilities. DGNB and similar certification schemes for buildings and construction make it possible to be certified to different levels of sustainability. DGNB can be seen as a translation of sustainability concepts or the SDGs into a building and construction context and includes the three areas: environmental, social, and economic qualities. In addition, there is a wide range of other concepts and models in areas such as LCA, circular economy, and various company certification schemes.

Thus, many good tools and ways of thinking about sustainability already exist. These tools make it possible to translate the original ideas from Gro Harlem Brundtland's definition of sustainability as meeting current needs without compromising the needs of future generations. However, it is a big and complex task, and it is essential that we all work seriously with its implications and make the effort to understand the principles and connections in order to apply it to our projects.

YOUR REFLECTIONS

In this book, we have included breaks for reflection called 'Your reflections' for all chapters. Reflection is a key capacity in sustainability. Without reflection, there is no learning and no implementation. If you want to make the most of your time with this book, make time for these sections. Make your own notebook and reflect with yourself on the implications of this knowledge for you. What will you change in your project and how?

You have likely already encountered most of these concepts before, but if you do not actively engage with them, it can be challenging to integrate them into your project practices. There is a tendency – for many different reasons – that sustainability is addressed in a somewhat superficial, checklist kind of way. But something else happens to the understanding if you work with them concretely in your own project context.

It can sometimes be difficult to get started with sustainability in a large project. It is common to feel the need to complete a project quickly, with each person confident in their specialisation, and eager to showcase their abilities. However, this can often lead to rushing ahead without careful consideration. It is necessary to slow these processes down and create space for dialogue on sustainability. It requires a certain degree of understanding of the basic principles and the ability to propose methods to work with. If you don't have an idea of the sustainability process, it will be very difficult to create any kind of sustainable innovation.

Engaging in reflection breaks and reflection work necessitates effort. It may feel like an interruption of the reading flow. Being able to break the flow to make it practical and applicable for yourself is a skill in itself. It is a skill that is important if you want to succeed with sustainability in your project management. It's about examining your own attitudes, thoughts, and wishes for your future as a sustainable project manager: how will you prove to yourself that you are making a difference in sustainability?

Therefore, take out your notebook and write down your thoughts, ideas, and reflections on the below questions.

How can you create your own sustainability experience privately before you do it professionally?

It is helpful to have personal experience on a new topic before you start managing other people. A basic hypothesis in leadership is you need to be able to lead yourself before you can lead others. So, consider what changes you can make in your personal life towards sustainability. One example would be to set a target for reducing your electricity consumption. Many energy companies have now made it possible to track your electricity consumption on an hourly basis. What do you use electricity for in your home and how much? Where does it make sense to change habits, and how consistent can you be in changing those habits? You will find out that it takes a great deal of attention and follow-up to create real improvement.

Which processes will you test in your projects?
There are different frameworks for including sustainability in projects. Are there any ideas from the presented suggestions that you would like to try to apply in your future work? Have you got your own ideas for good processes to increase sustainability in your project? Also, for your project, try to make a list of all the parameters you can influence that will increase sustainability – think in terms of environmental, social, and economic parameters. Which parameters can you control on your own and which ones do you need the agreement of the steering group or other stakeholders to influence?

You need to practice 'double vision'
Working with sustainability requires a systemic awareness of the overall sustainability impacts of the project, but it also requires targeting specific parameters. You need to pay attention so that you are not just repeating the old industrial behaviour in the decisions in the project. Have you thought about the required personal development of the project participants, do you have stress levels under control, should you take on apprentices, how do you work on building trust with suppliers and subcontractors, what about involving stakeholders such as environmental organisations, and so on. How do you ensure that you are both goal-oriented and aware of the systems in which you work?

READ MORE

Here are some suggestions for further study.

'Breaking Boundaries' by Owen Gaffney, Johan Rockström, *et al.* (2021) (Gaffney, 2021): this book is the starting point for our review of planetary boundaries. It describes very precisely and with data the planetary issues we are facing.

'The Climate Book' by Greta Thunberg (2022) reviews the scientific basis of the climate crisis, explaining how the climate crisis works, how it is changing the planet, and how it affects us all. The book also describes what we have done about it already and what we need to do now.

'Doughnut Economics – Seven Ways to Think Like a 21st Century Economist' by Kate Raworth (2018) provides a professional critique of the old economic thinking and a new way forward, underpinned by an economics that aims to keep us within planetary boundaries and above the social boundaries that must be minimum requirements. At https://doughnuteconomics.org/ you can find a range of tools related to this economic theory (doughnuteconomics, 2023).

The SDGs are systematically reviewed at: https://sdgs.un.org/goals with a review of all targets and indicators. Progress reports can be found at: https://www.un.org/en/our-work/support-sustainable-development-and-climate-action.

'The Age of Sustainable Development' by Jeffrey Sachs (2015) describes the data basis for the development of the SDGs and is worth reading to understand how the goals are linked.

European Water Stewardship proposes a certification with a specific focus on water stewardship: https://ews.info/.

ISO 14001 is the international standard for environmental management: https://www.iso.org/standard/60857.html.

Gaia: A New Look at Life on Earth by James Lovelock (Sachs, 2015), Operating Manual for Spaceship Earth by Buckminster Fuller (1969, 2008), and The Biosphere by Vladimir Vernadsky (2013) are interesting historical precursors to our current understanding of global sustainability.

Hyperobjects by Timothy Morton (2013), on the contrary, is a leading-edge philosophical book on environmental problems that proposes a new way of looking at pollution – as large hyper-objects that we as humans are inside and are part of. This offers a vision of the future in which we will have to live in a new world marked by what has already happened.

Other recommended books that provide in-depth understanding of the sustainability crisis and opportunities are: The Hidden Life of Trees by Peter Wohlleben (2018), The World is Blue by Sylvia E. Earle (2010), Breading Sweetgrass by Robin Wall Kimmerer (2020), The Sixth Extinction by Elizabeth Kolbert (2014), and The World Without Us by Allan Weisman (2022).

An interesting science fiction novel to read is 'The Ministry of the Future' by Kim Stanley Robinson (2021): it helps you imagining the future of the process of working for sustainability with stories from all over the world. It provides a comprehensive look at what the future decades may look like.

Earth can be viewed online from the ISS Space Station. It is said that most people who see the Earth from space are mesmerised by the overview effect. They suddenly realise how small and fragile the Earth is, and that there are no real borders between people, but that everything is interconnected: https://www.youtube.com/watch?v=_HGQZlK08gQ.

Part 1

Project Management is a Distinct Discipline

doi: 10.2166/9781789063349_0043

Chapter 1
Decisions create movement

It's not hard to make decisions if you know your values.

Roy Disney

Projects move when someone does something – when action is taken. Actions require decisions. Therefore, project movement depends on the project's overall ability to make decisions. Ensuring good decision-making processes is one of the key responsibilities of the project manager. Projects consist of thousands of decisions. Many of them are not even recognised as decisions but are simply choices we are hardly conscious about. The more consciously we make decisions, the greater the clarity, the more aware we are of our choices and the better the projects.

This chapter describes the overall framework for how to create a good structure for decision making in projects as a basis for effective management. How do I get off to a good start? What is the 'DNA' of the project and how should I use that knowledge? What is the project's framework or contract and what parameters should I manage by? What are the key awareness points for decision making? Answering these questions gives us key knowledge about the management structure of the project and ensures project agility.

The chapter provides a review of key overall principles and structure. A central and difficult competence is the ability to make decisions under uncertainty. It is a classic dilemma that everyone must inevitably confront and resolve in their own unique manner. Decisions under uncertainty challenge our desire to make 'safe decisions'. The chapter also includes and discusses the important concepts of the project triangle, project management models, and the follow-up parameters, which serve as a kind of generic map of a project's decision-making structure.

The steering committee plays a central role in the decision-making structure because it sets the overall framework and principles for all decisions, and it is here that the major fundamental decisions of the project are taken. Because the steering committee's decisions affect everything in the project, the steering committee is

Authors: Pernille Ingildsen and Mette Thorkilsen

a very important decision-making body during the lifetime of the project. The chapter concludes by offering valuable guidelines, tips, and strategies to foster effective collaboration between the project manager and the steering committee.

KNOW YOUR PROJECT

The word project is a compound of the two Latin words 'pro' meaning forward and 'ject' meaning throw. Thinking of a project as 'a throw forward' or a 'leap forward' is in many ways a good metaphor. Projects are about the future and have a sense of uncertainty about where the throw will land. When we think of throwing something forward, we have an idea of what we want to hit. But anyone who has taken part in competitions to throw stones at a target just 10–20 m away knows how difficult it is. We also have a clear sense that some have more flair for it than others, that training helps, that concentration helps, and that the way you think about the throw helps. It's the same in project management.

Projects come in many sizes and complexities. At one end, they can be hard to distinguish from what you would call a task. The classic example is to look at organising a party as a project with schedules, budgets, risk analyses, and so on. At the other end of the spectrum, projects can create things so new, different, and complex that they can change the world – the 1969 moon-landing project is a prominent example at the other end of the scale. In the last few decades, we have learnt a lot about how to make projects successful. We have realised that the tools and approaches required to succeed with a lunar-landing project can also improve the success rate of smaller projects.

A classic approach to categorising projects is to divide projects into C-projects, which are relatively simple, B-projects, which, because there are many of them and they contain a certain complexity, are the ones that typically take up the most bandwidth in an organisation. Finally, A-projects are particularly large and complex, and most organisations only carry this type of project out rarely. There are various definitions of 'what a project is' and how it differs from the more mundane 'tasks'. A good understanding however comes from the below list of the common attributes of project definitions (Olsson *et al.*, 2019):

- There is a well-defined start and end date – unlike operations that run day in and day out.
- A specific budget is allocated by the organisation to carry out the task.
- Cross-organisational or external competences are often needed.
- The focus will be on the time and budget allocated and the desired quality.
- Solutions are needed that are either new, appear in new contexts, are untested, complex, time-pressurised, or risky.
- It includes changes, new processes, or new products.
- An impact is created for a specific target group (stakeholders).

Construction projects will have different characteristics, for example, depending on the time of implementation, the client's needs, risks, importance, decision-making process, and so on. The characteristics of the project itself often set the framework for the choice of tender and contract types, forms of cooperation, choice of construction methods, necessary competences, user involvement, and so on. For example, complex projects with new and untested solutions will focus on developmental processes, whereas projects with low complexity and well-tested solutions will focus on small continuous improvements. To avoid time and financial pressures in the project, it is necessary to establish clear strategies for project execution and communicate these strategies to all parties on an ongoing basis. A wrong strategy can ruin the project before it has even started.

The characteristics or profile of the project depend on several factors such as:

- Scope of the project: duration, budget, quality level, risks, impact
- Importance of the project for the 'owner' or other actors
- Stakeholder overview of the project – how many, their influence, and importance
- Necessary resources and specific competences
- The level of awareness in the organisation of the task, solution, and methods
- Stability of the environment (political, economic, technological, regulatory, standards)
- Geography and culture

The complexity of the project can be determined based on the above factors and give the project manager an indication of how challenging it will be to implement and thus where the project manager should focus his/her attention. The project manager can benefit from a systematic assessment of the different parameters mentioned above to understand the profile of the project. If the project manager leans too heavily on own experience and perceives the current project as like his/her previous projects, the scope for solutions become limited and the project due to a lack of creativity may not have the desired effect. There is therefore a need for the project manager to take on new projects with an open mind – with a 'beginner's mind'.

All projects have their own profile with minor or major variations, and it is precisely here in the variations that the project manager's assessment of the project's profile can make a difference. It is not always the actual classification of the type of project that is important, but rather the discussions the project manager has with the project team or colleagues about the project and their perceptions of the project. The participants in the project will have different expectations and perceptions of the project and, not least, attitudes to the solutions. These opinions will characterise the participants' actions along the way.

Examples of descriptions of project complexity that can help to define the characteristics of the project, together with financial, time, and organisational conditions are shown in Table 2.

The project manager can get some help by answering the below 10 questions:

(1) Where is the gravity of your project? With the stakeholders? In the chosen solution (if it is defined)? In the risks? Or?
(2) What is the urgency to start/finalise the project?
(3) How tight is the time schedule?
(4) Is the project in a portfolio with other projects that can 'steal' attention from your project?
(5) Are there many different actors and a complex project organisation?

Table 2 Example of different parameters to assess the complexity of the project.

	Low	Medium	High
Technical 'gap'/ complexity	Existing methods – not complicated Tried and tested technology in a tried and tested set-up	Minor deviations from existing methods/processes	Difficult to describe scope/content and many obstacles. New technology or tried and tested technology in a new set-up
Risk profile	Uncomplicated, simple, and no obstacles	Challenging, but with some clear showstoppers	Difficult to define and many obstacles perceived as difficult
Knowledge of technology	Current technology	Customised technology	New technology
Environment, legislation, market, culture	Familiar environment, new authorisations not necessary	Different culture and market, but there is still an understanding of the legislation	Unknown environment and/or legislation Complex environmental permits needed

(6) Does the public have attention on the project? (a lot of communication)
(7) Have you chosen to introduce new forms of cooperation, technologies, or other innovative approaches in the project?
(8) Do you expect end-users to be involved in the project? How much? In which phases?
(9) Is the training of the end-user or other stakeholder groups a significant part of the solution?
(10) Will it be difficult to measure the impact of the project once it is finalised?

Understanding and having a sense of the project's profile is important from the very beginning of the project because we take several strategic decisions early in the process that will influence the project's progress and subsequent decisions. These initial decisions work best if aligned with an accurate understanding of the project's profile. A greater awareness of the project's profile from the start will make the project participants understand the 'nerve' of the project and ensure better decision making.

DECISION MAKING UNDER UNCERTAINTY

In projects, the project manager may not always possess the necessary knowledge to make well-informed decisions in a timely manner – even concerning important decisions. Some project managers may be almost paralysed by these situations. But it doesn't work. It is necessary to face this conundrum – to take responsibility, even when we feel we don't know enough. Some projects tend to fall into a trap of 'analysis paralysis', where the many options in a project are analysed for too long to find the best decision. Staying in this state for too long can be devastating for a project. On the contrary, you can't just work fully from your gut feeling either. Making decisions and taking chances are often required in the beginning of a project. However, when you recognise that this is an inherent circumstance in most projects, it often becomes easier to address and consequently progress the project despite it. One strategy is to make design decisions that permit adjustments at later stages as the project progresses and more information becomes available.

Especially at the beginning of the project, there will be a need for clarification and make fundamental decisions that will have an impact on the result – whether you have the necessary knowledge or not. This dilemma is illustrated in Figure 7. It demonstrates that crucial decisions must be made, despite the limited knowledge available at the beginning of the project. To overcome this problem, you need to use equal parts experience and intuition within your project team and consider the possibilities of changing decisions once you have learnt more down the road. There is a lot of talk these days about 'safe to try' decisions that help us overcome our tendency to avoid taking risks. 'Safe to try' decisions can be seen as experiments that have a low-risk profile and that make us wiser.

One way to overcome your internal decision-making barrier is to understand that making decisions requires more clarity than certainty. This means that there are many decisions where achieving certainty is very expensive and time consuming. But it is relatively cheap to establish a risk profile, consider the extent of the risks, and prepare a strategy for managing the risks. Risk analysis frequently enhances safety for all parties involved in uncertain situations, thereby facilitating consensus building within the project team and steering group.

When the project encounters major and important decisions, the project-steering committee is involved. Even if you do not have to make the major decisions yourself, you still have an important role to play, because it is your

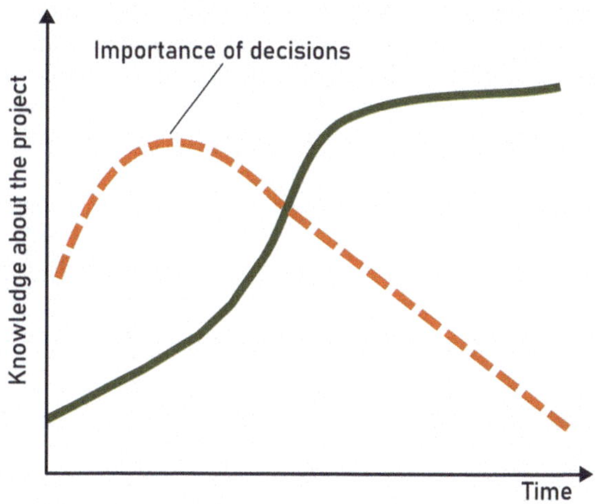

Figure 7 Illustration of the need to make decisions at the start of the project (important decisions fx) based on limited knowledge and high uncertainty.

task to provide the basis for decision making for the steering group by analysing and ensuring good planning of the project as early as possible. It is a basic challenge for all project managers to prepare and communicate the decision basis to the steering committee. This requires good communication skills and sometimes also some pressure to make the decision. The steering committee usually faces the same decision-making dilemma as you do. Therefore, the project will have much better conditions if you are able to create clarity for them when they must make decisions in an uncertain field.

It is beneficial to consider establishing the steering committee's trust in your abilities as a project manager as a distinct responsibility of yours. When the steering committee feels that you have their needs in mind, they will relax more and trust you when they sometimes need to take a leap of faith. Therefore, it is crucial that you constantly 'put money into their trust account'. You do this, among other things, by facilitating their decision-making dialogues in a good and impartial way. Also here, clarity is more important than certainty, and being able to communicate this clarity is a key competence for building trust, both in the major decisions in relation to the steering committee and in the minor decisions in relation to the project participants.

THE PROJECT TRIANGLE

The project triangle is a key concept to understand and always have in mind as a project manager. The project triangle shows the interdependence of time (the duration of the project), resources (the financial and human resources needed to complete the project and now also the Earth's resources), and quality (the main stakeholders' desire for the results and the value of these) (see Figure 8).

The project triangle is the starting point for discussion between the project owner and the project manager about what is possible for the project in terms of the balance between time, resources, and quality. The project triangle can be seen as a joint agreement or project contract between the project manager and the project owner. The corners of the project triangle must be defined, managed, and balanced by the project manager together with the project participants from beginning to end. Analysing and planning the project in the start-up phase and

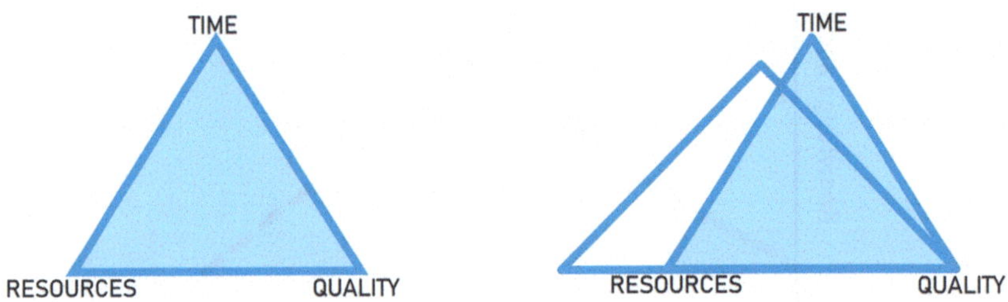

Figure 8 Corners of the project triangle, time, resources, and quality, represent the project framework. On the right, a project triangle under pressure in terms of time (less time for the task) and in need of more resources (alternatively, the quality level should have been adjusted) are shown.

defining the project triangle is an iterative process, where the project manager together with the project team discusses the different elements of the project. It may take some time; however, it is a good investment. When you manage to ensure that the project triangle is clear and understood by all parties you minimise friction, conflicts, and regrets in the project.

Project managers and the project team often find that the corners of the triangle will be squeezed and pushed from the beginning to the end of the project. The project environment, owners, users, and stakeholders have a tendency to want greater, cheaper, or faster results. Whichever corner is under pressure will always affect either one or both other corners. So, to ensure a successful project, the project manager must have continuous knowledge of the 'status' of the triangle, so that he or she can act wisely and clearly and manage the project triangle in a coherent way. It does not make sense to increase quality without adding more resources or ensure more time – similarly, it does not make sense to move the deadline without this also having consequences for resources or quality. Yet project managers often inadvertently agree to this in project negotiation situations without realising it. Slow down and never say 'yes' in the situation to please – you can often 'buy some time' and make the scope change decision to after you have analysed the situation. Discuss with the project team what is feasible. As a project manager, you need assistance in negotiating the project triangle with the project owner or steering committee.

If or when the corners of the triangle are squeezed, there are usually several possible options to handle it, and the chosen solution must generally be accepted by the steering committee. As long as you work within the project triangle, you have considerable freedom to execute the project. However, when encountering challenges related to the triangle, it's necessary to involve the steering committee. Preparing an overview of the project triangle requires thorough preparation, engaging dialogue, and relevant data.

The word responsibility is a conjunction of the words: 'response' and 'ability'. That means that the person who assumes a responsibility must have the ability to respond both in terms of competence and authority. Often responsibility and guilt are confused with each other and therefore people fear taking responsibility. Thus, it is beneficial to clearly define the concept of responsibility for oneself. It means that if something goes wrong you are the one to find the proper response. This also means that you ought to take great measures to pre-emptively avoid things going wrong – if not you probably owe somebody an explanation.

A central aspect of project responsibility lies within the project triangle. The essence of the project manager's responsibility to the project owner can be boiled down to having a clear overview of the project triangle and reacting in a skilful way to the pressures on the three parameters of the project triangle. A project manager who is new to the profession may have slightly skewed understanding of responsibility and believe that it means that you must fulfil all requirements to the project triangle no matter what happens. This can easily create stress and fear in the project manager, who has just embarked on a 'leap forward' and thus cannot anticipate everything that happens or everything that a project contains. This results in the sense that the project is an almost impossible task to succeed with – unless you make the budget of the project very large – which is rarely possible. And it is usually not even desirable to make the project's budget or timeframe very large. There is often a good point in keeping an 'appropriate pressure' on the project framework to ensure that the project moves forward.

The 'project responsibility' should hence be understood so that if something happens so that the framework does not hold, it is the project manager's responsibility to do something about it. Project managers 'owe answers' – not guilt.

Being able to effectively managing and openly communicating about the project triangle will enhance the trust of those around you. First and foremost, it helps you and the project owners to avoid unpleasant surprises – most often in the form of late-discovered budget overruns. In addition, clarity about the project triangle will make it easier for all participants to understand the range of outcomes available for decisions to be made in the project. Done in the right way, the project triangle can increase the possibility for innovation. It is therefore a good idea to talk to the project team about the project triangle and its state on a regular basis. Understanding the project triangle clarifies the space of freedom in the project for everyone, thus creating the right space for the creativity of the project participants.

At the conclusion of a project, the primary focus is on evaluating the project's worthiness in terms of investment (project triangle balance) and its success in generating the anticipated positive impact or benefits. The project's success often hinges on stakeholder satisfaction with the outcomes, making it crucial to manage expectations throughout the project's duration.

PROJECT MANAGEMENT MODELS

A company's quality management system is primarily established to ensure uniformity and clear structures for the execution of work in the line organisation. In the project world, there is a counterpart in the form of a company's project management model that describe the management structure, general principles for project implementation in different phases, tools, guidelines, templates, and so on.

The purposes of a project management model are:

- To ensure a common structure and understanding of the different phases of projects
- Providing project managers and participants with basic project management methods
- To ensure that all critical activities are carried out, none forgotten
- Ensure a shared language around project participants
- Provide clear and concise descriptions of the management structure, roles, responsibilities, and decision-making competences

A project management model incorporates defined phases that the company uses for various projects. These phases typically include idea, planning, design, execution, and realisation. They serve to break down the project into manageable components and provide a framework for specific tasks within each phase. Each phase includes descriptions of its purpose, outputs, required analyses, checklists, and applicable tools.

In addition, project management models should also include a description of the steering structure, also known as 'governance'. This typically encompasses roles and responsibilities for project participants, approval processes based on project type, decision-making authority, and meeting and reporting procedures. Project management models are usually documented in handbooks or digital solutions, outlining the implementation of project elements in different phases, as well as the governance structure regarding significant project decisions.

In general terms, the task of the project manager is to take the project through the tasks involved in each phase. Each phase ends with the project manager presenting the results to a steering committee, which then 'opens the gate' to the next phase. The framework for the overall project and for each phase is defined by the steering committee or, in practice, often in cooperation between the project manager and the steering committee. The framework includes definition of deliverables, success criteria, financial framework, and delimitation of the project against other parts of the organisation and other projects. As long as the project remains on track and adheres to the agreed framework, the project manager has full autonomy to guide the project through its individual phases (see an example in Figure 9).

The change of phases by going from one phase to the next is named gates and should be understood as a decision gate that opens access to the next phase. This 'opening of a gate' usually takes place at a steering committee meeting (sometimes called a gate meeting), if/when the project manager has given a clear overview of the results of the completed phase and the plans for the next one (or in other words, has a proposal for the framework of the next phase and can therefore visualise this framework). The steering committee meetings should be generic and include a look back at the finalised phase and its results and main decisions. At the gate meetings, the steering committee will need to accept the deliverables of the phase and there may also be several key decisions to be taken for the further progression of the project. In cases where there is a handover from one project manager to another, it might be beneficial for the departing project manager to showcase the outcomes of the prior phase. The new project manager on the contrary presents the plans for the next phase and the rest of the project, as well as the future framework of the project.

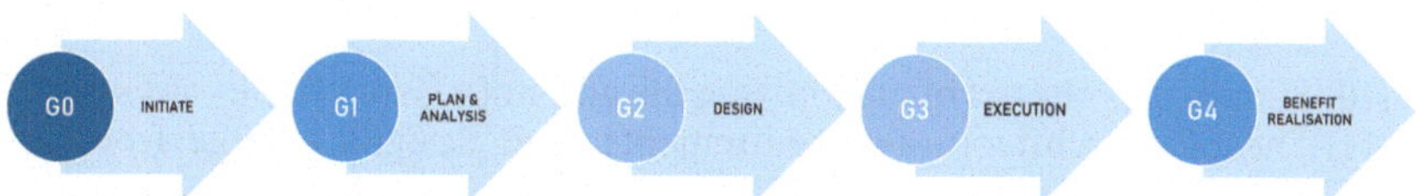

Figure 9 Example of a project management model with four phases and a realisation phase that focuses on the evaluation of the project and its results. The example is shown with gates, where G0 illustrates the decision to initiate the project and G4 marks the handover and initiating the evaluation of the project.

Interesting alternative models for executing projects can be found in the agile world, where SCRUM, Agile project management and Safe, for example, emerge – inspired from the lean approach. These are mainly built and adapted for projects in the IT world. In 2001, a group of IT developers wrote the concept of agile project management, which they called 'The Agile Manifesto'. The aim was to change the way companies manage projects and improve the way we manage and capitalise on the resources of IT developers and other project participants.

The Agile Manifesto reads as follows (Agilemanifesto, 2023):

> We are uncovering better ways of developing software by doing it and helping others do it. Through this work we have come to value:
>
> - Individuals and interactions over processes and tools
> - Working software over comprehensive documentation
> - Customer collaboration over contract negotiation
> - Responding to change over following a plan
>
> That is, while there is value in the items on the right, we value the items on the left more.

The Manifesto is underpinned by 12 principles and is considered a project management philosophy. The principles are:

(1) Our highest priority is to satisfy the customer through early and continuous delivery of valuable software.
(2) Welcome changing requirements, even late in development. Agile processes harness change for the customer's competitive advantage.
(3) Deliver working software frequently, from a couple of weeks to a couple of months, with a preference to the shorter timescale.
(4) Businesspeople and developers must work together daily throughout the project.
(5) Build projects around motivated individuals. Give them the environment and support they need and trust them to get the job done.
(6) The most efficient and effective method of conveying information to and within a development team is face-to-face conversation.
(7) Working software is the primary measure of progress.
(8) Agile processes promote sustainable development. The sponsors, developers, and users should be able to maintain a constant pace indefinitely.
(9) Continuous attention to technical excellence and good design enhances agility.
(10) Simplicity – the art of maximising the amount of work done – is essential.
(11) The best architectures, requirements, and designs emerge from self-organising teams.
(12) At regular intervals, the team reflects on how to become more effective, then tunes and adjusts its behaviour accordingly.

The principles are typically supported by several concepts, values, and tools that drive agile project management, such as 'sprint', 'the 5 roles', 'frontloading', 'burndown chart', and so on. The agile approach to project execution also uses concepts such as goal decomposition, estimation, and planning. There is a high degree of value creation integrated into the form and methods of the agile world, which is recognised from the lean world (Toyota's production system).

How you execute your project depends on the project type, scope, complexity, characteristics of the deliverables, and so on. Regardless of which model you use, the agile manifesto and principles are worth reflecting on, as they contain many of the ideas we know from the partnering mindset and lean construction concept (https://leanconstruction.org).

MONITORING AND FOLLOW-UP

Steering committee meetings hold significant value within a project, marking important milestones and enabling critical decision making. Regardless of their purpose, it is crucial to come prepared with a recommendation and a comprehensive overview of the project's finances, progress, and quality.

To gain an understanding of the project's current state, a well-structured management plan can prove invaluable. Collaboratively created by the project manager and the team, this plan incorporates regular inputs to track progress. The management plan encompasses multiple parameters such as time, resources, quality/deliverables, risks, and impact/benefits (evaluated through project success criteria). As such, key information related to progress and risks will be presented in the management plan:

- What is the indicator for our measurements? *Ex. milestones/top 5 risks*
- What is the necessary information? *Deadline for the milestones – The risk factor*
- How and by whom is data collected? *Ex. progress visualised in milestone plan/sand-visibility/impact diagram showing the status of risks (risk manager)*
- When: *Everyone in the team reports monthly for their own action area*

The financial aspect of the project receives significant attention and is the subject of many discussions and conflicts. The project triangle's corners represent time, quality, and resources, which all have a financial impact. Whether the project manager requires more time, additional manpower, increased product/workflow, or wishes to change the quality level (sustainable), it incurs a monetary cost. Managing project resources necessitates careful planning, diligence, courage, and decisiveness. The initial financial framework allocated to the project often consists of an estimate based on experience and preliminary project sketches. The project manager will gain knowledge through the initial iterations of the five planning tools described in the subsequent chapter, allowing for a more precise determination of project finances (time, quality, employees, sustainability, etc.) as a basis for ongoing monitoring and the management plan.

One approach to gaining an overview of the project's finances is implementing 'earned value management', which serves as the foundation for financial management within the American organisation Project Management Institute (PMI, 2021). Analysing the project's earned value at specific points in time reveals potential deviations from the expected cost and/or schedule. Projects consistently experience variations in the actual project cost, as the settlement of work performed occurs at different paces and involves shifts in the hierarchy among the various project parties. Thus, it's essential to consider potential delays in obtaining the accurate value of actual costs for monthly reporting within the company, depending on your role in the project. Evaluating the earned value in construction provides a more useful indication of the resources utilised (man-hours, materials, machinery, etc.) and their corresponding value on any given day (PMI, 2021).

Key figures using the methodology are illustrated in Figure 10:

PV	Planned value	Budget allocated for planned work.
EV	Earned value	Work carried out in relation to approved budget.
AC	Actual costs	Actual costs incurred for work performed.
EAC	Estimate at completion	Expected total cost of project.

The most common analyses look at the cost variance, budget variance, and schedule variance to get indications of the most appropriate actions to get the project's economy back on track. At the same time, calculations also provide a

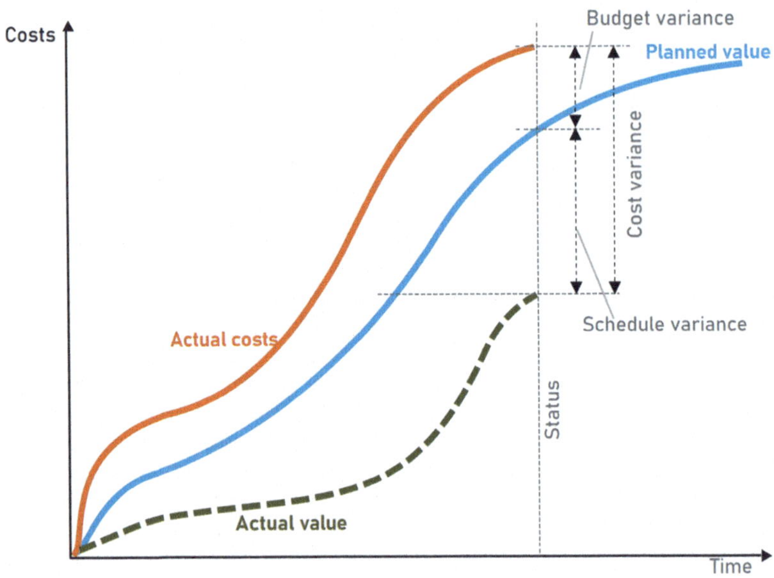

Figure 10 Illustration of key figures for the monthly financial/timely statement of the project.

picture of where the challenges lie in the project: do we need more time, more money, or are the deliveries challenged in terms of quality or quantity?

CV = Cost variance = EV – AC = earned value – actual costs
BV = Budget variance = PV – AC = planned value – actual costs
SV = Schedule variance = EV – PV = earned value – planned value

In the event of persistent deviations from the planned value, measures must be taken to bring the final estimate in line with the expected value. The planned value curve will largely be equal to the project budget, and the final estimate will therefore be an indication of whether the project's economy can be kept within budget. It is recommended to report deviations, even if it leads to variations in the final estimate within the regular monthly report.

It is natural for the final estimate to be regularly updated and change throughout the project - execution of projects is rarely linear. If there is a continuous deviation from the planned value and budget, it usually gives rise to discussions in the steering committee.

Good practices to ensure the financial viability of the project:

- No hidden reserves in the budget to ensure success - you need to be transparent, and the money could have been spent on other projects if it had not been tied up in your project's budget.
- Use an estimation tool to detail activities and their costs (e.g., three-point estimation).
- Risk analysis makes the basis for assessing the sum to prevent and handle risks.
- The summary of contracted expenses and final estimates play a crucial role in managing the project triangle. Updated project estimates are provided monthly.
- Design and technical solution should be handled by people with the right competences - especially in the execution phase - so that the cost estimates and levels are as accurate as possible.
- Regularly assess the status to accurately forecast and identify trends with an appropriate level of detail.

However, even with a good and inclusive framework and continuous follow-up and management of the project, the framework will sometimes be challenged. For example, there may be a choice between an expensive solution that is better than a cheap solution. Or that two roads ahead both have risks associated with them, and it is difficult to assess which one is worse. Or even if you take the path with the least risk, there is still a substantial residual risk that the steering committee must be aware of and accept. Or external demands put the framework under pressure - should we keep the framework and reject the demand or expand the framework and accommodate the demand?

Other examples of challenges to the framework are:

- Cost increases or delays in relation to materials
- Technical solutions cannot work as intended
- Suppliers/contractors have a shortage of resources/labour

- Change in regulatory requirements
- Design changes

Often, the changing elements are already known from the risk analysis. The preventive measures may not have been strong enough, or a single incident occurs that causes the project framework to slip. This again emphasises the importance of the project manager having an overview of time, budget and spending, quality, risks, and the expected results of the project. The steering committee must be always informed about the risk profile as a basis for their decisions. Regardless of the extent to which the steering committee is aware of the events that cause the framework to slip, it is the project manager's task to present the problem to the steering committee with a recommendation for a solution to the problem. In some cases, changing the framework will take time and until a new framework/project triangle is approved, the project manager will have to continue the project in the old framework – with an eye on the expected new framework. It may be necessary to stop some activities and withhold payments until the clarification is finalised. Such difficulties may happen once or even several times in a project's lifetime.

Decisions concerning changes to the project framework, the handling of external issues, or fundamental design choices are taken at the steering committee meetings. The steering committee requires a clear description of the issue and suggested solutions, along with alternative options, as soon as possible. This ensures that committee members have ample time to prepare effectively. The project manager has the task of presenting a basis for good decision making including the possible solutions and their consequences. It is a good practice for the project manager to propose a preferred solution based on his/her own understanding of the situation and the success criteria. Still the material needs to be clearly unbiased.

Occasionally, a project may encounter challenging situations that require escalating decisions to the steering group. These procedures should be outlined in the organisation's governance structure or project management model. This becomes crucial when the agreed framework cannot be met or appears likely to be exceeded in the future. Additionally, serious risks or unforeseen situations can necessitate such discussions. In these critical situations, it is vital for the project manager to engage with the project owner to determine the appropriate course of action. Work should continue within the established framework until a decision is made by either the project owner or the top management/board of directors. Through collaborative problem solving and acceptance, the project owner shares responsibility for the associated risks. It is the project owner's duty to be informed when the framework is challenged at any level.

STEERING COMMITTEE WORK

The work with the steering committee may be challenging. In some projects, there is no steering committee or project owner. In other projects, there is a steering committee with members who are inexperienced in steering committee work. Clear expectations and dialogue on steering committee work often end up being part of the project work. Project owner and project manager are linked and should be appointed for all projects. The steering committee meetings act as a link to ensure that the project and the rest of the organisation are synchronised, and expectations are aligned.

A project is a temporary organisational structure and often lives a kind of 'parallel life' in relation to the rest of the organisation. It is therefore important that there is a structure for regular meetings between the two structures of organisation to ensure cohesion. In a way, the project manager and the project owner have two different perspectives on the project. For the project manager, prioritising the progress and success of the project is essential; however, for the project owner, the focus is how the project fits into his overall understanding of all the organisation's needs. For the two to work and fit together, there needs to be clear communication and mutual respect for the two different perspectives.

During the execution of a project the project owner generally acts as a 'customer' in a customer-supplier kind of relationship. The project owner is the customer who orders a (complex) deliverable from the supplier (project manager). Whether the steering committee is established with three to five members or the steering committee consists of one person, for example your supervisor or line manager (they will subsequently become the project owner), there will always be a certain expectation that the project manager will deliver what the customer/project owner wants. There is a need for the 'customer' to be able to communicate their wishes to their 'supplier' in relation to, for example, performance, finances, time, features, and so on. The customer also needs to be able to follow and participate in the necessary decisions along the way.

The project manager represents the project owner in the project and is responsible for ensuring the project's progress and success. To effectively navigate the project landscape, the project manager should have the necessary authority, which can be defined through the project triangle framework. Negotiating and defining this framework is a joint responsibility of both the project manager and the project owner. Additionally, a document describing the scope of the project can be used to clarify what project includes and excludes.

The steering committee, in a way, functions like a board of directors for the project 'company'. It is essential for the relationship between the project manager and the steering committee to follow best practices, just like in a director–board relationship.

To ensure timely decision making and uphold the project manager's authority, regular meetings between the project owner and project manager are important. Prior to a steering committee meeting, a pre-meeting can be conducted with the project owner to allow the project manager to raise issues and receive feedback in advance. This ensures that the project owner is informed and supportive. However, the steering committee meeting remains important as it provides an opportunity for all members to address important questions and issues, ensuring that decisions are made with qualified input and respect for the organisation and the entire project. Therefore, both the project owner and project manager hold the responsibility of fostering a safe forum where the project can be discussed and where all parties are heard and understood.

It's important for both the project manager and the steering committee/ project owner to understand that the purpose of the meetings is to monitor project progress and make decisions. Additionally, the steering committee members should provide the necessary resources for the project and act as project ambassadors. For the project manager, the steering committee meeting, especially the preparations leading up to it, offers an opportunity to regain a managerial overview of the project. It can be challenging to fully assess the project's status when immersed in the day-to-day operations. Therefore, the preparations for steering committee meeting can be used as an opportunity for reflection on the project's progress and to determine any necessary adjustments in project management.

For example:

- Are the costs and timetable on track?
- Are there any unaddressed risks?
- Which success criteria have we overlooked?
- Do we need to make any adjustments to deliveries?

Many steering committee meetings become less productive when there are misunderstandings in communication, which may easily create unnecessary conflicts. To maintain a constructive atmosphere, it is essential to strike a balance between listening and speaking.

It is not uncommon for participants in such meetings to become engrossed in the details, as we often believe that 'the devil is in the detail'. However, this focus on minutiae can hinder important discussions from gaining a higher/broader perspective. Higher-level decisions are often more intricate and carry greater risks, which may make individuals unconsciously prefer the comfort of delving into details. This pattern can pose a lot of problems. The issue is that strategic decisions are not being made or are being postponed due to time constraints in committee meetings. It is therefore crucial to address any unspoken concerns, even if it may be uncomfortable and both parties risk appearing bewildered. It is important to make the project owner aware that details should not be the main concern of steering group meetings and will be handled by the project manager and the project team.

All steering committee members should prepare by reading through the material, but in many organisations, you can't always be sure of this. Therefore, present your material as if one or more participants are not fully prepared or familiar with the material – they may even have read it but misunderstood parts of it. This can be frustrating but accept that people's brains work differently. Consider the scope and format of the material, aim at a length of the material, so that people can prepare in 5–20 min.

When a steering committee meeting goes well, it can give the members a sense of calm, allowing them to step back from the project. If they trust the project manager, they can concentrate on other tasks, allowing the project manager to work without micromanagement.

All decisions and their background should be summarised in the minutes. It is also helpful to have an overall summary of previous major steering committee decisions in a presentation that can be referenced when necessary. Reminding the steering committee of previous decisions may be helpful and may prove valuable in situations where past decisions need to be reassessed. Additionally, there are instances where a decision should have been changed, but someone is unaware that the 'old' decision still holds significant influence.

Five tips for project managers on steering committee work and meetings:

(1) Build a mutually trusting relationship with your project owner.
(2) Never surprise the project owner and steering committee at the meetings – difficult messages may be wise to handle in dialogue prior to the meetings.
(3) Send out meeting material in good time.
(4) Ensure that mandated steering committee members are present for decision making.
(5) Be honest, clear, and flexible (if you are asked questions you cannot answer, you can always provide answers, analyses, etc. afterwards).

Good advice for the steering committee and the project owner is about acting wisely, calmly, and empathetically. The following can serve as a checklist for the steering committee as well as the project managers (who sometimes must 'educate' the steering committee, as not all have learnt how to act as steering committee members).

- Be aware that this is a meeting where the project manager may feel vulnerable, like an exam. However, it is important to remember that this is a collaborative effort and not a test – so do not attack.
- Projects consist of numerous details, and the project manager cannot be expected to have all the answers at hand.
- Steering committee members offer different perspectives, and what may be clear to them may not be obvious to the project manager. Therefore, it is crucial to listen carefully and provide precise information.
- The steering committee should consider the long-term implications and determine whether they need to make decisions that go against the project manager's wishes. If such a situation arises, it is important to ensure that the project manager understands the reasoning behind the decision. Otherwise, the project manager may feel compelled to follow arbitrary rules, which is undesirable.
- Express appreciation for the work involved in the project, recognising that you may only be seeing a small portion of the effort.
- The project manager requires a clear framework and well-defined decisions to navigate unforeseen events effectively. Thus, it is essential to provide clear mandates and decisions.
- Instead of offering solutions to problems, focus on articulating your success criteria and important needs. Allow the project manager and the team to come up with suitable solutions that align with the project and their working methods.
- If your speech is unclear or confused, the project manager may rely on body language, subtle remarks, and previous statements to guess your intentions. This can lead to misunderstandings, so feel free to ask questions if you need clarification.

- Rather than criticising, communicate your needs for reassurance, clarity, security, overview, understanding, and decision making.
- The project itself is already complex. Therefore, introducing additional layers of 'interesting innovative ideas' can jeopardise the success of the project. It is often better to keep things simple and understandable for everyone rather than striving for perfection.
- Be aware if the stakeholders are becoming difficult or have withdrawn from the project.
- Ask questions like: Is it still possible for the project to achieve the expected impact?

MAIN POINTS

The project manager's early-stage understanding of the project's characteristics is crucial in determining project-related decisions and their execution. It is valuable to engage in a dialogue with all project participants and collectively identify the project's key aspects. This will ensure that the decision-making processes are aligned right from the beginning.

The success of projects hinges on clear and transparent decision-making processes, even when faced with uncertainty. Prioritising clarity and a comprehensive perspective are crucial to handle the problem of limited knowledge. Understanding the project triangle – highlighting the relationship between quality, time, and resources – ensures consistency throughout the project. However, the project triangle faces constant pressure from the demand for more, cheaper, and faster results. As a result, it's the project manager's duty to maintain balance within this triangle.

Project governance structures should be apparent to all participants, with well-defined roles and responsibilities. Establishing a transparent framework and employing project management models streamlines the project's day-to-day execution. Project managers must actively manage resources, basing decisions on solid financial grounds and effectively conveying them to stakeholders.

Project managers must assure seamless decision making by differentiating between decisions that fit within the agreed-upon framework and can be taken by the team and those that necessitate clarification from project owners or steering committees. Productive steering committee meetings cultivate trust among all parties involved and deter excessive micromanagement by the committee itself.

YOUR REFLECTIONS

Understanding something 'roughly and unconsciously' differs greatly from knowing it after consciously reflecting on relevant issues. Gaining insights through the reflection process can bring clarity, and involving project participants in this process can provide even better input. Engaging in reflective dialogue about the elements discussed in this chapter may yield surprising and valuable information. It's often astounding how differently things are perceived and this can be seen and understood when collaborating using methods for collaborative reflection. Therefore, don't hesitate to include team members or project partners in the reflection process outlined in this chapter.

Which areas are new to you and which elements do you recognise from your everyday life?
Did anything surprise you or did your awareness become sharper while working on this chapter? Have you identified any skills that need improvement? How do you plan to continue developing them? What practice opportunities are accessible to you, and how can you obtain valuable feedback? Who in your network can offer helpful advice or support?

In what way is there clarity for all project stakeholders about the decision-making processes in your current project?
As you can read in this section, the overall responsibility of the project manager is to ensure the fluidity of the project, and this is facilitated by actively clarifying the project in a way that makes decision-making processes flow smoothly. Lack of knowledge about the project, unclear decision-making processes, and poor communication about the status of the project are often reasons for delays. How do you ensure that you bring the best available knowledge into the decision-making process? It is often necessary to make decisions without a fully informed basis, but how do you ensure that the knowledge that is present in the 'project network' is fully utilised?

Do you know your organisation's project management model (if you have one)?
The lack of a project management model creates ambiguity about project execution, misunderstandings, and unclear expectations within the organisation. Is there clarity in your organisation about how you execute projects? In what way can this be improved? In what ways are there clear or unclear guidelines on how to deal with different sustainability issues?

How do you manage the project?
A systematic approach to managing the project against a clear set of parameters makes it possible to facilitate decision making. How do you follow-up and manage the project? Do you have an overview of your project? What is the status of the project triangle parameters?

How does your steering committee work and what characterises your cooperation?
We often hear that the project manager's experience of cooperation with the steering committee resembles an examination. This is part of the old mindset. In the sustainable project management paradigm, the project manager and the steering committee are on the same team. By changing your mindset and approach to the steering committee meetings this way, will you be able to reach a more collaborative content in the steering committee meetings? It is key that you are prepared and communicate clearly about the project terrain. Based on this understanding, are there good questions you can prepare that create dialogue about the needs of the organisation and project owners – to steer clear of the examination feeling?

How can you introduce more sustainability into decision-making processes? How do you include sustainability issues in decision-making processes? Be aware that implementing sustainability in decision-making processes requires extra time, focus on change, and a new mindset. Where can you start and what does it take to succeed? How will you introduce sustainability into the decision-making processes? Is there anyone who can help you in your close or distant network?

READ MORE

Power in projects, programmes, and portfolios by John Ryding Olsson, Mette Lindegaard Attrup, and Niels Ahrengot (Olsson *et al.*, 2019) describes the basic elements, tools, and experiences in project management, projects, and portfolio management.

On the PMI website https://www.pmi.org/ (Project Management Institute), you can find information on certification, membership, standards, and publications on project work, training courses, and so on.

The Project Management Body of Knowledge (PMBOK), the Project Management Institute's (PMI®) project management standard – is a handbook that describes PMI's approach to project work in terms of processes and knowledge areas (PMI, 2021).

https://ipma.world/ (International Project Management Association (IPMA)). On IPMA's website, you can find information about certifications, knowledge, training, and so on (IPMA, 2022a).

Individual competence baseline for project management (www.IPMA.world). The tool is used either as a self-assessment to target competence development or in connection with preparations for IPMA certification. The book can be purchased at IPMA.world.

You can read more about Scrum (SCRUM, 2023) at www.scrum.org and learn more about agile project management at: https://ipma.world/?s=agile+project+management or in IPMA reference guide ICB4 in an agile world.

doi: 10.2166/9781789063349_0063

Chapter 2

The five essential tools

It takes half your life before you discover life is a do-it-yourself project.

Napoleon Hill

This chapter introduces the five most essential project planning tools that will significantly simplify your project management process and help keep it on track. Although there are numerous tools available for project management, mastering these five will provide a solid foundation for streamlined execution.

If you are like most project managers you have probably already been on a course and learnt about the five project planning tools, but perhaps you have not implemented them properly in your work. There are many explanations and excuses for the lack of implementation in practice. Many of the tools are used somewhat subconsciously: maybe it's because it wasn't necessary to be explicit; maybe you feel that you have it all in your head; you didn't quite understand the point of using the tools; your manager didn't ask for it, so you just never got started, or something else entirely. But what is it that prevents us from having the stamina and self-discipline to be much more precise and explicit in our use of project management tools? Try to think through the consequences of not implementing the tools. It's not just yourself who suffers. Use this chapter to give yourself a new opportunity to realise the importance, the interconnectedness of the tools, and finally to incorporate them into your project management practice.

When we ask project managers, 'What is required to create successful projects?', their answers are always more or less the same:

- Support from management and/or steering committee.
- Clearly defined objectives in terms of purpose, success criteria, and deliverables.
- Planning of the project is carried out jointly and the plan is actively used to monitor the progress of the project.
- The right and sufficient resources are present in the project.
- Stakeholders and management are engaged and involved along the way.
- A clear division of responsibilities has been established.

- Risks are managed proactively.
- There must be a good and committed cooperation in the project team.
- The project is widely and openly communicated.
- Project participants are active and there is a systematic follow-up.

To succeed with these primary needs, you need to master the five essential project management tools. The five tools will take you a long way in creating the foundation for success in the project; both for yourself and for everyone involved. The use of the five planning tools creates a simplicity and overview for you and the team, and a checklist for whether the project is 'in control'. The results of working with the tools can also be used in communication with the steering committee, internal and external stakeholders, and as onboarding of new team members.

This chapter introduces the five tools that have been tried and tested over many years (Olsson *et al.*, 2019). The language around the tools is clear and makes it easy for you as a project manager to involve your project team and even external partners in the work. The good start-up of projects goes through the use of the five tools shown in Figure 11.

Applying the tools at the start of the project will create a common process, commitment, and a well-defined foundation for the project work at hand. The five tools are process-orientated, and it is the project manager who initiates and facilitates the use of the tools together with the project team. The tools must be revisited on an ongoing basis throughout the project. The result of the

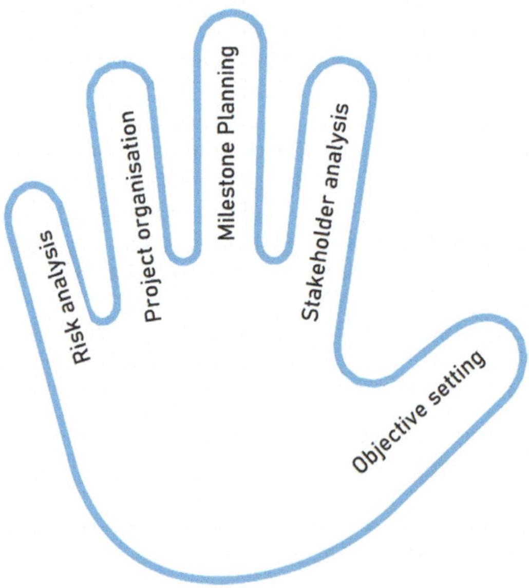

Figure 11 Five basic tools of project management are: objective setting, stakeholder analysis, milestone planning, project organisation, and risk analysis.

work will bring knowledge about the project into play that is necessary to make decisions – at the start as well as along the way.

Applying the five essential tools will give you and the project team:

- A planning basis to manage and monitor the project on an ongoing basis.
- Knowledge you can use in your daily communication.
- Shared vision and direction for the project.
- Natural processes for conducting the dialogue within the project team and stakeholders.
- Clear project team commitment.
- A shared understanding of project priorities.
- Discussion of difficult topics and thus better alignment.

Explicitly and actively using the five tools at the start and continuously throughout the project will be a great help for everyone who are touched by the project. With these tools, the project becomes transparent which means that everyone can contribute in active and valuable ways.

TOOL NO. 1: PROJECT OBJECTIVES

A clear formulation of the project objectives from the beginning has the following benefits for the project and its participants:

- Clear goals serve as a potent motivator for most project participants, keeping them engaged in their daily tasks.
- As a project manager, you'll quickly appreciate the value of having well-defined and precise goals once the project commences to avoid confusion.
- Objectives ensure a clear scoping of the project.
- Objectives are at the centre of defining plans.
- Objectives make it easy to communicate about the project to all stakeholders and ensure that they have the same understanding and recognition of the importance of the project.

Any serious project manager should take the task of formulating objectives seriously to avoid negative consequences when the objective is either too unclear, non-communicable, or simply forgotten!

Missing or unclear objectives make it harder for the project team to navigate as the project progresses. Everyone needs a clear 'beacon' to aim for. A lack of objectives makes it difficult to follow-up on whether the project is meeting its objectives. This can lead to many difficult discussions along the way or even worse, at the end of the project. Projects that fail often have inadequate or unclearly formulated objectives.

The objectives of the project must first and foremost reflect the needs of the 'customer' (internal or external), and this is the first thing to be worked on in the project. This work may well have been preceded by various analyses, problem-solving processes, or studies of users' needs, which have led to the start of work on the formulation of objectives.

In the 1990s, The Danish Minister of Transport was asked in an interview why we should build a fixed connection between the eastern and western parts of Denmark as a bridge connection across a belt (today known as the Great Belt Connection). The Minister replied: 'so that the people of the eastern part of Denmark can get fresh milk from the western part of Denmark'. If the project's sole purpose had been to transport fresh milk, there were various other, more cost-effective alternatives than constructing a billion-dollar bridge. Besides there had never been a lack of fresh milk in the eastern part of Denmark, so it was wrong in so many ways. In projects, unclear objectives create messy and unclear communication – and this bubbles all the way to the top.

However, we can learn two lessons from the story of milk and the Great Belt Bridge:

(1) We typically form our own very personal ideas and images of what a project is about from very few words or a broad and unclear formulation. Unclear wording and lack of dialogue on objectives often lead to misconceptions and myths about the real purpose of projects.
(2) When it is possible to create clear problem formulations, it is sometimes possible to find cheaper and/or better solutions for the project than the originally formulated solutions. In this way, clear problem formulations open the possibility of new thinking and innovation.

The three concepts of an objective

Effective objectives have three key elements and are formulated as shown in Figure 12:

(1) Purpose: Why are we doing this?
(2) Success criteria: How do we know if the purpose has been achieved?
(3) Deliverables: What have we created when the project is finished?

The formulation of the objective thus answers what is to be realised and why and how it should fit with the clients need. When we decide to launch a project, we are faced with a challenge to be solved, a fire to be extinguished, or

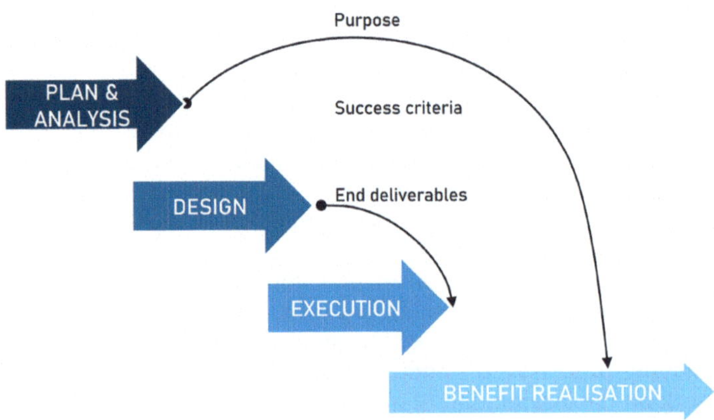

Figure 12 Formulation of purpose, success criteria, and deliverables.

an idea to be realised. The objectives serve as the foundation and initial step for working on the other four tools. Take your time to carefully craft the proper wording for the purpose, success criteria, and deliverables. Poorly formulated objectives can result in confusion, redundancy, or negatively affect the project's overall impact.

The objectives can be seen as a hierarchy. The client's objectives are the top objectives. The partners deconstruct these objectives and restructure them to suit their portion of the deliverable, ensuring coherence and relevance to their tasks. For instance, a contractor might be in charge of one or two deliverables. Consequently, the consultants, the contractor's project manager, and their team undertake the responsibility of refining the client's objectives into pertinent objectives aligned with their own deliverables.

Purpose of the project

The purpose serves as an overarching vision that guides the project towards the goals in the future. The purpose answers the question: when the project is finished, what have we achieved? What will be the benefits?

Comprehending the client company's strategy is crucial for aligning the project's purpose with its value creation and ongoing activities within the organisation. This strategic alignment also seeks to balance the project's ambition with the available resources that the company possesses and can allocate to the project.

Furthermore, it is essential to articulate the project's purpose when communicating with stakeholders about upcoming developments. The purpose generally includes words that describe the changes the project aims to achieve, such as 'increase', 'sustain', 'improve', or 'create'.

Purpose statements are often relatively short. A good way to find the right wording for the purpose is to answer the question: 'Why do we need to realise this project'? Examples of purpose statements are:

- To secure the future water supply in the city
- To expand the wastewater treatment plant by 20%
- To improve the ecological quality of a river
- To optimise the system to reduce leaks of laughter gas and other climate gases

The objective should be clearly defined in order to act as a shared reference point, guiding projects towards the achievement of desired outcomes and value.

Success criteria

The success criteria should clearly describe the project's intended impact, measured by quantifiable results. Evaluating the project after completion, these success criteria allow stakeholders to determine whether the project has met expectations and can be considered successful. Success criteria are directly linked to the project's purpose, with their formulation often stemming from organisation's fundamental values, and possibly its targets or key performance indicators.

It can be helpful to use the SMART rule to define success criteria:

S: is specific described?
M: is it precise and measurable?
A: can the criteria be accepted by key stakeholders?
R: is it relevant but also ambitious?
T: is it time-bound?

Example of objectives and success criteria:

Purpose: To establish separate sewerage with cloudburst protection in a neighbourhood.

Success criteria:

- Reduce the load from untreated wastewater overflows by 90% after a 3-month commissioning period.
- Detention basins provide a delay of at least 2 h for wastewater during cloudbursts.

Purpose: To renovate an existing building to make it more energy efficient.

Success criteria:

- Reduce electricity consumption by 20% and heat consumption by 30% compared to the existing building – 3 months after commissioning.
- 80% sustainable materials must be used for renovation.
- The building received an indoor climate label at the time of commissioning.

In the final assessment of the project, evaluating the success criteria is crucial. Thus, continuous efforts must be made to guarantee that the success criteria documentation is ready for the final evaluation.

If the success criteria for a project are too ambitious, it may become too challenging to achieve success, and participants can easily lose motivation. Conversely, if the success criteria lack ambition, project participants might not take it seriously enough and may even fail to reach the low ambition objective. Clearly, the success criteria should align with the financial resources available for the project.

Deliverables

The third and final aspect of formulating objectives pertains to the project's deliverables. Deliverables refer to the tangible, concrete outcomes of the project which may be visible, touchable, and audible and must be completed by the project's end date. To determine the deliverables, ask: what needs to be in place to achieve the purpose and the success criteria?

It is not always the case that all deliverables are perfectly defined according to the SMART rule at the beginning of the project. That's OK. The crucial aspect is ensuring these matters are resolved and clearly outlined well before the project enters its execution phase.

In the construction industry, project specifications are often developed from the start and finalised before the tendering process or contract signing with the contractor. In some cases, contractors are invited to participate in the design phase and help define the deliverables. Precise wording for final deliverables is crucial to avoid misunderstandings and ensure a specific quality level. This stage is also important for incorporating sustainability parameters.

Inadequate deliverable formulations may result in incomplete work and uncertainty about the project's intended outcome. On the contrary, creating deliverables that don't support the project's purpose wastes time. It's crucial to avoid formulating unnecessary deliverables, even if they sound appealing. Deliverables lacking relevance to the project's purpose only incur additional costs without achieving desired effects.

Formulating deliverables involves asking questions like 'how do we do it?' or 'what will it take?' with reference to the project's purpose. To ensure quality, ask why a specific deliverable is needed; the answer should be found within the project's purpose. A clear hierarchy between purpose and deliverables is essential for achieving desired outcomes.

Well-formulated deliverables are the starting point for the subsequent planning work. Set aside time to create good, solid formulations that are agreed upon by the project team – also use the SMART rule when formulating deliverables and formulate the deliverables in the past tense as if the work has been done – this makes it easier to follow-up later.

Examples of deliverables:

Purpose: To establish a separate sewerage system in a neighbourhood.

Success criteria:

- Reduce the load from untreated wastewater overflows by 90% after a 3-month commissioning period.
- Detention basins provide a delay of at least 3 h for wastewater during cloudbursts.

Deliverables:

- Two hundred metres of wastewater pipelines have been installed and commissioned.
- Two hundred and ten metres of existing lines in poor condition are lined and commissioned.
- Two rainwater basins with a volume of 2000 m^3 have been constructed.
- Operational staff are trained to operate the new sewerage system.
- Operation and maintenance manuals and new processes are documented.
- Communication about the project to customer has been carried out.

Purpose: To renovate an existing building to make it more energy efficient.

Success criteria:

- Reduce electricity consumption by 20% and heat consumption by 30%.
- Ensure that the renovation is carried out with the least possible inconvenience to residents (evaluated by a survey 3 months after completion).
- The project must be completed by the agreed date.

Deliverables:

- The building is retrofitted with 100 mm of sustainable insulation material.

- All windows and doors have been replaced with three-layer glass and a *U*-value of 0.8.
- A connection to the district heating network has been established.
- Plumbing and heating systems are in place.
- All appliances have been replaced with A++ models.
- Operational staff are trained in energy optimisation during operation.
- Operation and maintenance manuals and new processes are documented.

At the beginning of the project, the client typically outlines general final deliverables. As the project progresses, more specific partial deliverables are defined, which are included in the corresponding contracts with partners and suppliers. To prevent misunderstandings and conflicts about what needs to be delivered, it is crucial to clearly specify and align expectations at all stages of the project.

The process of goal setting

The objectives provide a unified direction and focus for the project team. The project objectives need to obtain approval from the steering committee before initiating the project. Clear communication between the project manager and the steering committee helps prevent delays and cost overruns.

Collaboratively developed by the project team, objectives take advantage of diverse knowledge and perspectives. Through dialogue, team members establish a precise and shared understanding of the project's goals. Although differing opinions can make these discussions challenging, they are crucial for refining and clarifying the objectives. The project manager must ensure that these conversations are comprehensive yet focused, ultimately leading to a well-defined vision.

Investing in goal-setting dialogue cultivates a mutual understanding and ambition for the project, laying the groundwork for future progress. This process necessitates individuals to embrace uncertainty and trust that the team will find a solution despite potential challenges. At times, explicitly expressing this belief may be necessary, especially when starting with limited or uncertain information. The essence of formulating objectives lies in believing that success is possible through the combined efforts of the project team.

Construction projects are often characterised by an intensive development phase in the early stages, where the project's fundamental values and objectives are gradually established, assumptions are identified, and ideas are developed. In the construction industry, the footwork of formulating the overall objectives of the project usually lies with the client. It is easy to forget to repeat this discussion later in the process when consultants and contractors are involved in the project.

Similarly, it is not enough for the sales and tendering department at the contractors to know the client's objectives for the project. All actors working on the project must know the overall objectives of the project and their own company-based objectives. If project managers and participants from consultants or contractors are not given the opportunity to understand and discuss the shared picture of the client's needs, there is a high risk that misunderstandings and conflicts will arise sooner or later in the project. Missing or inadequate objectives are equivalent to working without direction and meaning to the tasks. This leads to decreasing engagement and general confusion. This can be difficult to empathise with as a client who has already worked for a long time and already thoroughly understands the project's objectives. That's why it takes care, discipline, and patience to get everyone on board. This applies to all the project managers and participants who are 'on board' during the project.

Construction projects often run with different actors in more parallel processes at the same time. Be careful to ensure that the objectives also make sense in the organisational culture of the cooperation partner. This part is perhaps the most difficult, and the amount of sustained effort needed to ensure a shared image and direction for all project participants is often underestimated.

The scope of the project must be clearly defined in terms of objectives, that is, what is included and what is excluded from the project. A lack of definition is the source of many misunderstandings in all subsequent phases of the project. Going forward, the project objectives will be used for decision making, planning and, not least, to motivate project participants and key stakeholders in the project.

The project objectives are formulated as one of the first things in the project but can be subject to continuous adjustments along the way. It is quite natural that the project's objectives may be a little 'vague' at the start, and in some cases, it takes time to get the objectives to stand out clearly with all three components: purpose, success criteria, and deliverables. But it gets better as more knowledge is actively brought into play through dialogue.

Changing basic objectives during the implementation phase itself is often a bad idea and may easily cause several cascading problems. If necessary, consider whether it is better to recommend the steering committee to close/terminate the project and start a new one. Alternatively, consider how much change in objectives is necessary and then how much the overall plan needs to be corrected. It is essential that all those affected by changes are clearly informed about the change, its cause, and consequences. The project becomes almost impossible to manage if different participants have different perceptions of the objectives. Make sure the change is understood by all the partners of the project.

TOOL NO. 2: STAKEHOLDER MANAGEMENT

The stakeholders of a project are groups or individuals who have an interest in the project, influence the project, or are affected by the project. Especially the latter will not always be enthusiastic about the project. Resistance to the project can feel like it is coming out of the blue if you have not thoroughly mapped the stakeholders of the project. Dissatisfied stakeholders can cause major problems for projects and colour perceptions of the success of the project. Hence, this is an area that needs to be taken seriously regardless of your role.

At the centre of the stakeholder analysis is the question: who are we building for? Who are we creating the project for? We typically call them the users or key stakeholders. Fortunately, today's users are more involved than in the past and it requires more resources and more time for the project manager and project team to carry out user involvement processes. User involvement is initiated in the early stages of the project and key stakeholders are often to be kept in the loop throughout the project. This involvement adds great 'user-value' due to customisation to the users in terms of ease of use, health, and safety and other – often very specific – wishes and requirements from the users. In addition to the users, there are usually an additional number of other stakeholders that need to be managed continuously throughout the life of the project.

Stakeholders are managed according to their attitude – positive or negative. It would be better that all project parties know and understand the stakeholders, their needs, and interests. Hence, it can be a good idea if both consultants and contractors to engage with future users and consider their perspectives. The better everyone knows the stakeholders, the easier it is to deliver the right product with the right functionality; the best result – whether you are the client, consultant, or contractor. Throughout the project, stakeholders can be seen as project advisors, judges, facilitators, resources, opponents, and so on. Effective stakeholder management entails understanding the needs and perspectives of different parties, regardless of whether you agree with them or not.

Stakeholders can have different roles in the project. Some want to help, and others just need to be informed. Yet, others need to be involved and some need to be influenced. Some are involved from the start and others only in parts of the project's lifetime. Although numerous stakeholders may exist, only a select few hold significant importance. It is essential to allocate time and resources specifically for these key individuals. The stakeholder analysis is used to identify, prioritise, and manage the most important stakeholders. Typically, a stakeholder analysis is carried out during the planning phase of the project. Later, the analysis is repeated regularly throughout the project to ensure a good stable stakeholder involvement.

The three steps of stakeholder analysis

The purpose of carrying out the stakeholder analysis is to ensure satisfied stakeholders and a successful project in their eyes. Another good reason to conduct the stakeholder analysis is to establish valuable knowledge about the project by getting stakeholders inputs and understand their perspectives on the project. Thus, it is crucial to engage in open stakeholder dialogue involving your project team, as it fosters a shared understanding of the project's nature and its surrounding stakeholders. Moreover, the stakeholder analysis lays the foundation for a comprehensive communication plan, which encompasses all stakeholders, clarifies objectives, and offers valuable input for organising the project.

Stakeholders can be curious, happy, angry, ignorant, engaged and, not least, focused on their own needs. All of which we can handle with the right attitude and investment. Stakeholders often enquire about the project's purpose, its impact, success criteria, and desired outcomes. Many of these questions are addressed through the formulation of objectives. However, stakeholders may also bring up important issues that the project team has not previously considered or discussed, further contributing to the project's development.

Stakeholder analysis is a three-step process with a few simple rules:

(1) Identification of stakeholders
(2) Analysing and prioritising stakeholders
(3) Planning specific activities for handling the selected stakeholders

Step 1: Identification of stakeholders

The identification of stakeholders can be carried out as a brainstorm in the project team so that everyone can contribute. Project participants will often have different experiences or perceptions of why and how stakeholders can affect the project. A brainstorm creates energy and engagement if it is organised and facilitated properly. It can be done quietly, noisily, creatively, systematically, or whatever you want to stimulate.

Start by writing each stakeholder on a post-it note – be as specific as possible.

Stakeholders are all those who influence the project (may be involved in the process) and those whom the project affects (must live with the outcome). Typically, stakeholders are relevant to consider at various stages of a project. Generally, most stakeholders have a keen interest in the project deliverables. However, those who influence and show interest in the project itself are often also interested in its progress and content throughout the process.

Stakeholders can include clients, end-users, and contractors to suppliers, authorities, neighbours, and competitors. It's helpful to view stakeholders as individuals or small groups because perceptions of a project can vary significantly even within one stakeholder organisation. Stakeholders often have diverse perspectives on a project and may require different approaches based on their unique viewpoints.

Step 2: Stakeholder analysis and prioritisation

To get an overview of all the stakeholders listed, put post-it on a wall or board and place them in an 'Attitude vs. Importance' chart. 'Attitude' shows how you think the stakeholders perceive the project, and 'Importance' is your judgement of how important they are to the project formally or informally. Under importance, it also matters whether the stakeholder has informal power or influence over the project's outcome or deliverables (see Figure 13).

The project team might not always agree on interpreting stakeholders' needs and interests; however, engaging in discussions is vital for incorporating diverse perspectives and valuable insights. Therefore, when moving through the second step, ensure a conducive environment that allows differing opinions about stakeholders and simultaneously encourages everyone to contribute their expertise and knowledge.

It is uncommon for a project to have the resources for intensive stakeholder management. Therefore, prioritise which stakeholders you want to manage and which ones to brief, monitor, and engage with occasionally. Focus on stakeholders with a negative attitude and/or high importance. Some may have low importance and high negativity, but are managed for various reasons, whereas others have high importance and a positive attitude, such as management or steering committees. Prioritisation is based on the diagram

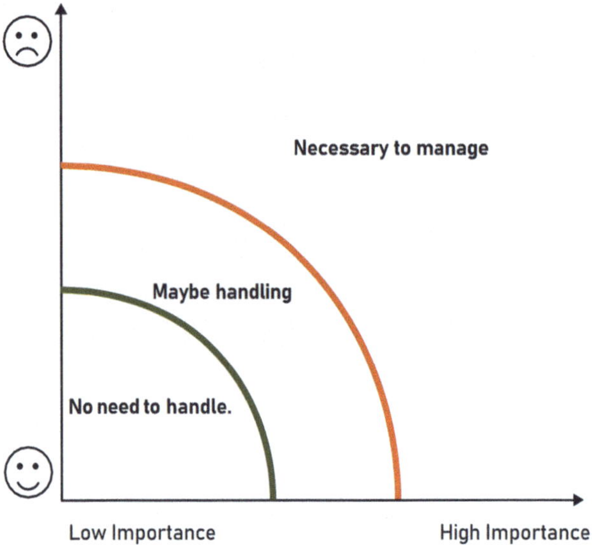

Figure 13 Diagram to apply prioritisation of stakeholders according to their importance and attitude.

along with personal experience, knowledge of specific stakeholders, and the project's structure.

Once you have prioritised the stakeholder, the next step of the analysis is to look at the project through the stakeholder's eyes – sometimes with their help. Conversations with stakeholders will often reveal interests you hadn't anticipated in advance – and interests that sometimes seem illogical (from your point of view). Ask if possible – prepare good questions that make it possible to learn something new about the stakeholders.

One focus of this step should be to identify what stakeholders see as the pros and cons of the project. Stakeholders are typically concerned with: what does it mean to me? What can I use it for? Where do I need to change? And why? What will be helpful for me during the project and regarding the deliverables? For certain stakeholders, conducting interviews based on these questions might be beneficial. For others, it's essential to understand the various perspectives they may have regarding the project's processes and outcomes. As you practice empathy, your ability to view the project through the eyes of different stakeholders will steadily improve.

Most project managers and participants are (naturally) centred in their own image of the project and may find it difficult to imagine or even accept other people's perspectives on the project. This requires openness. Instead of seeing stakeholders as a burden, it is often helpful to try to think of them as a potentially valuable resource that you can benefit from on an ongoing basis, and by whose influence and input you may increase the overall value and quality of the entire project.

If, after the analysis, you find yourself unsure about the opinions and attitudes of some stakeholders, it is essential to meet with them and initiate a dialogue regarding the project. This ensures that you have the best possible starting point and helps to prevent unexpected issues later in the project. Misjudging a stakeholder's attitude, opinion, or importance early on can lead to increased costs, delays, or compromised quality later.

Step 3: Planning specific activities for dealing with selected stakeholders

In the third step, discuss and decide on activities to address the stakeholder's issues effectively. For many stakeholders, it's possible to change their attitude towards the project, particularly if their mindset is based on incorrect assumptions or preconditions. Clearing misunderstandings or misinterpretations through meetings or providing accurate written material is crucial. In some instances, stakeholders may possess valuable knowledge or solutions that can benefit the project. Involving positive stakeholders as ambassadors can also greatly contribute to the project's success by influencing other stakeholders.

Ensure that you clearly describe the specific initiatives, such as meetings, workshops, and newsletters, tailored for each stakeholder. The outcome of this phase plays a crucial role in developing a comprehensive communication plan for the project, outlining the responsibilities of each party in terms of communication. Include these activities in the project's timeline to maintain progress.

The stakeholder analysis is regularly updated or revisited as required, because the landscape may evolve over time. Changes in attitudes, introduction of new stakeholders, and other factors necessitate ongoing adaptation of the analysis by the team to ensure timely detection of any shifts in the overall situation.

An example of a stakeholder management plan is provided in Table 3.

Specifically on user involvement

Users or key stakeholders can be invited to participate in the project by involving them in relevant fora, meetings, or workshops. The involvement of users should be structured in such a way as to capitalise on the knowledge and experience they represent. Individual users typically lack decision-making authority in projects, so it's crucial to establish the extent of their influence and the appropriate timing for their involvement. Clarity regarding what can and cannot be influenced is also essential. Thus, a key responsibility for the client's project manager is to define and structure user involvement. This process includes prioritising needs and desires in relation to financial constraints and project goals and clearly communicating these factors to users. User groups often possess limited knowledge of construction and technical details, which can create challenges in conveying the project effectively.

It is primarily the client who interacts with users and is responsible for effectively communicating technical projects to them. After this interaction, the insights gained from users must be relayed to other stakeholders in the project organisation. A crucial task for the client's project manager is to present a narrative that best represents the needs of most users. The project's success

Table 3 Example of a stakeholder management plan.

Stakeholder	Attitude/ Importance	Stakeholder Benefits	Stakeholder Disadvantages	Activity	Time/ Responsibility
Future user	High	Canteen synergy by sitting together	Long commute, new colleagues, smaller office	Involvement in user workshops	Early in the project cycle – ongoing follow-up/ project manager
Neighbours	Between	Can keep up with construction	Noise from construction	Informed regularly via flyers	Whole project process/ project leader
...	...				

relies heavily on ensuring that user insights are not lost during this process and are shared with the relevant parties involved.

The project manager acts as an 'interpreter' and mediator between users and the rest of the project organisation and must consider the various concerns of the end-user. The wishes or requirements of users must be weighed up against the consequences and presented to the decision maker.

TOOL NO. 3: THE MILESTONE PLAN

With the objectives and the stakeholder management plan in place, the next exercise is to draw up an overall timetable – a milestone plan. The plan contains the most important milestones (~30–35 according to the overview) in a temporal aspect.

Eisenhower once said: 'plans are worthless, but planning is indispensable' and there is some truth in that – because the process of making a good plan is important in creating an understanding of the project's execution. Often, the plan will change as the project progresses, but adapting the plan is much easier when you have an already coherent plan describing the entire project (in various details).

Two errors in particular are often seen in relation to scheduling. These errors reduce the effectiveness of the timetable. The errors are:

Error 1: In a classic execution-focused culture planning is often the very first tool to be applied. Often the steps of getting a clear understanding of objectives and stakeholder analysis are typically skipped. The consequence is a kind of 'blind planning'. The objectives must be the starting point for the milestone plan, and they need to be logically broken down into milestones. If the objectives are missing, it usually results in a plan with no link to success criteria and deliverables, which results in an unfocused and ineffective plan. What is needed is a robust, coherent, and clear plan that provides an overview for everyone in the project. That's why the preparatory work is key.

Error 2: Planning starts by listing activities and placing them one after the other. Starting by creating a milestone plan instead is much better because it ensures that activities are coordinated accordingly. The formulation of the activities between the milestones is only done once the project organisation is fully in place. By that time, a person responsible for each action area or selected milestones will have been appointed. They will be personally responsible and have the authority of identifying the necessary activities to fit the overall milestone plan. Therefore, you should prepare the milestone plan first and stick to it as the overall structure.

The milestone plan has the following objectives:

- To provide an overview of the critical milestones and deadlines in the timetable.
- To help create clear roles and responsibilities.
- To motivate project participants.
- Provide a good way of communicating to key stakeholders.
- Used to manage and follow-up on the individual milestones.

The joint milestone plan, utilised by the client, consultants, and contractors, is based on the client's objectives. The plan must be cohesive, encompassing the entire project across all levels.

The creation of the milestone plan is carried out in four steps:

(1) Breakdown into work streams based on deliverables.
(2) Identification of overall milestones within each work stream.
(3) Discussion of dependencies of milestones between different work streams.
(4) Discussion of the timeframes and the division of the project into main phases.

If difficulties arise in agreeing on milestones or deadlines, it's likely due to imprecise objectives that require revisiting. Utilising the five essential tools involves an iterative process, and adjustments to previous work may be necessary. It's not uncommon to move back and forth between one or more of the tools several times.

Define work streams

The overview describes a plan based on work streams, which typically serve as headings for one or more deliverables. Dividing the milestone plan into work streams creates a clear and robust outline, enabling realistic planning. Work streams are identified according to deliverables and can be defined by disciplines, critical areas, or other factors. These areas are named to reflect the nature of the work involved. Each work stream comprises a series of milestones or activities that ultimately lead to one or more final deliverables.

Examples of work streams could be: 'outdoor activities', 'building closure', and 'installation'. The work streams can also be geographical, for example 'section A/B', 'area A', or similar. The most important thing when defining work streams is to create a milestone plan that is coherent, optimal, and provides a meaningful overview. Often a specific work stream is developed with a manageable number of milestones which eases the communication with stakeholders.

Identification of overall milestones

Milestones are defined as 'A planned state signalling the completion of one or more activities'. A milestone has no extension in time and does not in itself imply an action. Milestones are to be understood as a 'flag in the plan' signalling that an unambiguous state has been achieved. This could be, for example, a partial delivery or a decision that has been taken. Milestones can be effectively defined using the SMART rule. The wording should be as specific and measurable as possible, and in the present tense so that they are unambiguous and at the same time easy to follow-up.

Examples of milestones:

- The steering committee's decision on the design proposal has been taken.
- All pipes are fitted and sealed.
- Cables are laid for part A of the building.
- Filter XX is installed.

Discussion of dependencies between work streams and milestones

After formulating the milestones and determining the appropriate sequence within each work stream, the next step entails focusing on dependencies between the various work streams and specific milestones. These dependencies can be illustrated using arrows connecting milestones across work streams. The milestone plan then transforms into a comprehensive process plan, allowing for a clear understanding of the project's progression within each work stream, as well as its dependencies and potential realisation timeline.

Discussion of the temporal sequence

The final step involves estimating the project's timeframe and ensuring that the milestones are entirely SMART, including a specific date for each milestone. Once the dates are set, the initial draft of the milestone plan is prepared. This draft must be further refined by outlining the activities required to accomplish the milestones. After completing the risk analysis, it may prompt another update of the plan, incorporating preventive measures from the risk management strategy. A milestone plan is shown in Figure 14.

Milestone plans rarely change a lot, whereas activity plans may change many times during the project. Don't focus too much on detailing the plan beyond the first 3–5 months. Instead, develop a comprehensive schedule that clearly outlines dependencies, which holds greater importance. Effective planning involves breaking down the project from a strategic level to individual tasks, highlighting their purpose, deliverables, milestones, activities, and completion.

In addition to milestone plans, there are plans such as tendering plans, contract plans, process plans, and weekly plans, all based on and referencing back to the milestone plan.

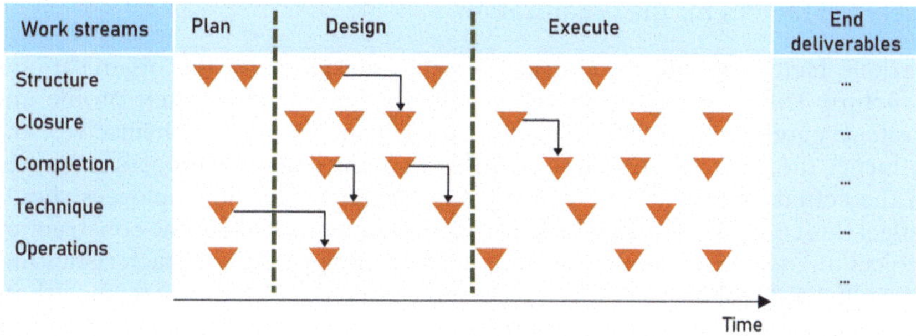

Figure 14 Milestone planning: (1) define work streams, (2) formulate milestones, (3) adjust milestones according to dependencies, and (4) insert time and possibly phasing.

In a typical construction project, numerous interdependencies exist, making a comprehensive process plan essential. These interdependencies between various disciplines often result in conflicts, necessitating their prediction and coordination. To address this, coordination is generally ensured at both an overarching level and through weekly plans that focus on specific aspects.

An important planning tool for the construction site is Lean construction, which was developed in the 1990s, inspired by the Japanese Lean Production system. Lean construction is about creating value for the customer by eliminating waste and activities that do not create value as much as possible (Liker, 2021). Lean construction is an effective initiative that aims to create improved planning, seamless workflow on construction sites, and enhanced employee well-being.

Thus, there can be many different plans with different purposes. The overall aim for all plans is to meet the milestones in the milestone plan. The milestone plan should guarantee that the project will work out in the end despite all the complexities.

TOOL NO. 4: PROJECT ORGANISATION

The project truly gains momentum when it is well-organised, and the project team is assembled. In many cases, each project brings together new faces, with only a few familiar ones. It may sometimes feel like everyone is starting from square one with each new project, but this is very often the nature of project work.

Many companies lack clear or sometimes even existent organisation for their projects. It is crucial to establish a clear understanding of the thought process behind the chosen working methods. How are the project participants selected for each case? Is it up to individual managers to make these decisions? Does the company have a professional approach to organising and staffing? In order to ensure the success of the project, it is imperative to define roles and responsibilities and create an organisational chart.

Essential factors for the organisation

Various factors should be considered when deciding on the organisational structure. Different organisations or actors should consider their profile and strategy carefully in relation to how they want to present themselves and influence the projects, this might be different in the tendering process compared to the actual execution. The collective competences of the employees should reflect this consideration. It is crucial that the company's role in the construction project aligns with competences that support the project's characteristics and are sufficiently robust to ensure successful completion. Fielding a team without the necessary skills is a sure path to failure, resulting in problems and financial losses. Relying solely on a single project manager's competence is unrealistic and risky. Therefore, it is important to consider how to ensure the resilience of the full team. Too many project teams are fragile and dependent on individual efforts, which is detrimental to the industry. It leads to employee exhaustion, financial losses, ineffective projects, and missed opportunities for creating significant value.

When setting up the project organisation, it is important to consider the structure and people carefully so that resources, skills, and disciplines are appropriate to the nature of the project. In some projects, the set-up is straightforward, and it is easy to staff the project. In other projects, stakeholder management can be a challenge and therefore requires different competences in the project team. In other projects, the chosen solutions are new and untested, while the stakeholder picture is well known. Therefore, projects must be organised with care and according to the understanding of the individual project.

All projects need an agile and flexible organisation that is adapted to the nature of the project (solutions, stakeholders, time factor, risk factor, importance, etc.). This may mean that not all project participants are involved from start to end or that the organisation and staffing change along the way. Changes in the team organisations are common during many projects, especially in larger projects, which makes team management more challenging. Thinking of organisation as something that is somewhat dynamic is helpful in the process of dynamically assigning the right people with the right competences at the right time.

The client function is considered the 'owner' in terms of legislation on agreements with consultants and contractors. Often, the term client encompasses an internal organisation that is much more complex, consisting of all parties on the 'owner's side' (e.g., owner, developer, investor, user, specialists, and building administration/business manager). As a client, it's important to keep the role as the owner in mind, regardless of how the organisation is structured and which part you represent. Towards authorities, delivery systems, and society, the client has a responsibility that needs to be represented effectively, and which the underlying organisation must ensure.

In the client's organisation, the line management often chooses the project manager and possibly the project team. The project owner could be the line manager or the responsible person in the organisation. If the organisation has a formal project management model, it may indicate who is responsible for developing the project idea and who is subsequently appointed to manage the project. Regardless of the practice in your organisation, it is crucial to understand the various roles and their significance throughout the project's lifespan.

Roles and responsibilities

Project work differs from traditional manufacturing companies in terms of its organisational structure and employee experience. Unlike fixed structures with well-defined procedures and roles, project organisations operate without a conventional management hierarchy. Instead, the project manager assumes leadership over participants but usually lacks formal employment powers. This often requires borrowing staff from the company's line organisations, where employees have their own managers. Therefore, it is crucial to clearly define 'roles and responsibilities' to avoid any confusion or ambiguity.

Most projects benefit from the following standard project organisation:

- Steering committee (decision-making group)
- Project manager and project team (work unit)
- Reference groups (entities that assess partial solutions, give feedback, and/or provide knowledge or acceptance)

In Figure 15, it is shown how to organise the project by splitting the different roles into a decision-making structure.

The steering committee

The project owner (or project sponsor) 'owns' the project and must ensure that the project creates value for the organisation. The project owner is typically also the chairperson of the steering committee and decides, together with the project manager, when and how the steering committee meetings will be organised.

The steering committee should be composed of representatives who support the final deliverables of the project. Steering committee members can be changed along the way so that they constantly reflect the stage and nature of the project. The main work of the steering committee is to make the necessary

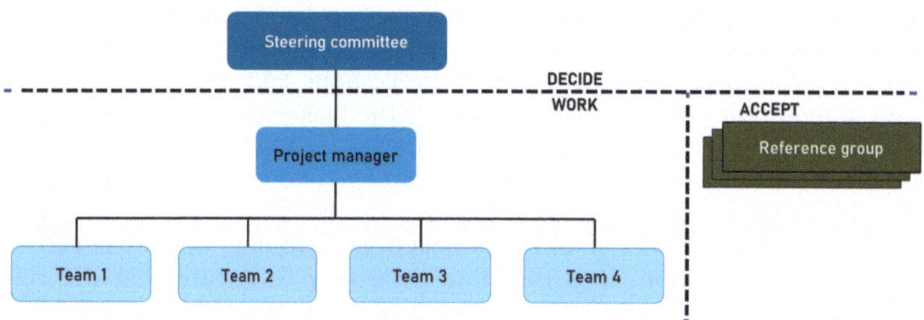

Figure 15 Roles and decision-making structure of the project organisation.

decisions and secure the right resources. In addition, they must follow-up on the progress of the project and act as ambassadors for the project. The role of an ambassador involves the members of the steering committee promoting the project to the outside world by communicating positively and ensuring a sustained marketing of the project. The size of the steering committee should not be too large (preferably only three to four people), as this makes it easier to reach decisions.

In construction projects, the role of the project owner typically lies with the client, who initiates the project and invests in the project. Generally, there are no representatives from consultants and contractors in the steering committee. Partnering projects, public–private partnerships, and strategic partnerships are an exception to this, where it is more common to have a mixed steering committee representing the different actors on the project – or both having a client steering group and a partnering steering group with all parties.

The type of contract also plays a role. In a large project with many specialised contractors, it may be difficult to invite all parties to the steering committee. In such cases, it will often be the construction manager at the construction meetings, representing the client, who will act as the 'steering committee' at the operational level. In other types of contracts, such as turn-key contracts, main contracts, or large-scale contracts, it is possible to set-up a steering committee with representatives of all the main actors together with the client. This takes time and should be seen as an investment to ensure good cooperation in the project.

Every project requires thorough consideration on the best set-up of the steering group or groups. The main task of steering groups is to take strategic decisions. One way to organise is to consider what kind of strategic decisions will need to be taken during the lifespan of the project and who will be responsible for which decision and then organise accordingly.

Project manager and project team

The role of the project manager is primarily to lead the project team and ensure that the objectives are realised. In addition, the project manager will plan, organise, follow-up, inform, communicate, report, and coordinate. The project manager has a project team with members who help to fulfil the tasks of the project. It is necessary for the project manager to accept a high degree of delegation and therefore work with confidence that the project team can carry out the work as agreed.

The project team identifies tasks, their content and expected time, and resource requirements, and then carries out the tasks within the agreed time. Project team members need to be solution-orientated and able to implement the solutions. So here, too, there must be a wide range of competences.

In the construction industry, most projects involve multiple companies working together. It is crucial for the project organisation to operate effectively inside this collaboration. It is therefore necessary to ensure that the right competences are in place, that reporting and decision-making processes are well defined and that everyone has a common understanding of who does what (roles and responsibilities).

It can be challenging when core members of the project organisation come from different companies with their own business objectives, different circumstances, and a different understanding of the project. It cannot be emphasised enough how important it is to involve and create cohesion and a shared vision of the project. This also means that each company sends a project manager into the project, and therefore there are usually several people with the title of project manager on the same project. However, there can only be one project manager reporting to the steering committee.

It is important to describe the project organisation as accurately as possible so that everyone can navigate as an effective part of the team. If you are unsure about the right composition of your project team, it is always a good idea to look at the milestone plan, which sets out the main work streams. When the responsibility for work streams is robustly staffed, the organisation is generally robust.

Good questions to ask when setting up project teams with internal as well as external staff are:

- Does the team member have the right technical competences?
- Does the team member have the necessary experience in the key tasks of the project?
- Is the team member motivated for the task and is time available to fulfil the job?

When preparing the project organisation, it can be helpful to work out 'optimal competence profiles' for the necessary roles and descriptions such as 'roles and responsibilities' before the project team is staffed. If you as the project manager have an overview from the start, it is easier to staff the group and thus ensure 'the right person on the post'.

Reference groups

In order to create transparency towards the main stakeholders, the project can be organised with one or more reference groups that can provide input to the project. The reference groups are experts, advisors, or end-users who provide input or are consulted in relation to different solution proposals or processes. When deciding which reference groups an organisation should have, it is important to determine which of the stakeholders should be included and which should be handled in different ways. It is also a good idea to be clear about which topics should be discussed at the reference group meetings. Remember that reference groups do not generally approve solutions or similar, they only have an advising role.

TOOL NO. 5: MANAGING RISKS

Experience shows that most risks are predictable. Therefore, there is no way around systematic risk analysis and follow-up. Risk analysis is probably one of the most valuable times to spend in projects – it has a huge pay-back. Yet it happens all too often that risk analysis is not carried out in projects – at great cost.

Many project managers believe – and even say – that they have all the risks in their heads. But this approach has a whole range of problems:

- The risk picture is based only on your own knowledge.
- The risk picture is not known by the team and stakeholders.
- The respective risks are not objectively assessed and prioritised.
- There is no systematic management of project risks.
- There is no checklist to see that all relevant risks are addressed in the project plans.
- It is difficult to involve the steering committee in the most important risks.

There will always be risks involved in projects, such as dealing with unknown knowledge or processes, new suppliers, and methods. Effective project management involves preparing a risk profile for the project early on, preferably before the project is approved. This risk profile then serves as the foundation for subsequent strategies in design, tendering, contract form, and more.

The purpose of risk analysis is to anticipate and address negative events that may occur during a project. It involves identifying and mitigating risks, as well as preparing for their management if they cannot be avoided. By conducting risk analysis, the project manager and team can prioritise their response to these events and minimise the impact of potential disasters. Additionally, risk analysis stimulates open communication within the project, allowing for the integration of new knowledge and experiences. It is worth noting that risk analysis may also uncover positive events or opportunities that can enhance the project's outcome.

Working with risk analysis provides everybody with a better compass to navigate the project safely. It should, therefore, be a required routine activity and a systematic discipline.

Definitions and concepts: and the three steps

Risks are defined as negative events that can occur with a given probability. A risk always has a cause and a consequence. It is important to be aware of the difference between cause and consequence. For example, too many analyses have noted 'delay', as a possible unforeseen event. It is not. Delay is a consequence of an unforeseen event. An unforeseen event can be, for example:

- Key person disappears from the project.
- Fracture in concrete element.
- Shortage of raw materials.
- Stakeholder changes position on an important issue.

The consequences of unforeseen events are, for example, delays, cost increases, lack of deliverables, decreased quality, and solutions not working optimally; anything that prevents or reduces the success of the project. The cause is what leads to the unforeseen event, and there can be several causes leading to the same risks. In the example of 'fracture in concrete element', it could be due to errors in set-up, poor concrete quality, or something else. Prevention activities need to be linked to each cause.

The risk overview including the most important risks should be presented to the decision makers in the steering committee to ensure they are aware of the investment's risk profile, including the associated risk costs and benefits. It is crucial that the project manager does not bear the risk knowledge alone. The steering committee should be actively involved in continuously updating and monitoring the risk situation. By involving the steering committee, stress levels can be reduced for everyone involved. Additionally, there may be several risks that require judgement in their management, and it is advisable to engage the steering committee rather than making those decisions alone.

The risk analysis needs to be updated regularly, at a minimum every time the project changes its phase and when major changes happen in the project. The reason for the update of risk profile is to ensure that the current risk management plan is up-to-date, clear, and well prioritised.

Step 1: Identification of risks

Identifying risks is best done through a brainstorm where each participant contributes based on experience, knowledge from other projects, and gut feeling. Prepare a checklist of the 'normal' general types of risk from similar projects. Some project managers have their own checklist created based on experience. This can be used to ensure that the brainstorm has been fully covered. Checklists complement but do not replace the work of identifying risks in the new context of the new project. Each project has its own individual risk and new risks can always arise. It is crucial to pinpoint them as early as possible, ideally at the beginning of the project.

A risk checklist may include the following topics:

- Definition of the objectives and scope of the project
- Design parameters – users' needs and interests
- Geological or other physical conditions (e.g., no outside areas for the construction site)
- Access and traffic management in relation to the construction site
- Quality of dialogue with users on needs and interests
- Stakeholders in the neighbourhood
- Political conditions
- The relationship of the different parties with the authorities
- New laws and regulations
- Price development of raw materials, products (e.g., as a result of general development, local level, currency conditions)
- Structure and resources of the project organisation – key persons
- The chosen form of procurement and contract
- Dependencies on other projects nearby or in the portfolio
- Errors in design/execution

- Project changes along the way – especially in the execution phase
- Lack of building materials/resource availability
- Occupational health and safety and environmental conditions – sustainability profile
- Theft and vandalism from the construction site
- Tight timetable
- Lack of communication from some or all parties
- Unclear contractual conditions or missing descriptions
- Unclear or missing service descriptions (incomprehensible/quality)

It is advisable that everybody starts with brainstorming by themselves and then share their ideas, rather than performing the brainstorm all together – this will help avoid group think.

Step 2: Analysis and prioritisation

For each risk, the cause and consequence are described in words. To indicate the level of probability and consequence, a scale of 1–5 or 1–10 can be used and the total risk (risk factor) equals the probability multiplied by the consequence (see Figure 16). The array of risks can then be prioritised according to the highest risk factor. Spend most time on the most significant risks (those with the highest risk factor). Discuss and decide whether the different risks should be accepted, transferred, mitigated, or eliminated and then describe the relevant

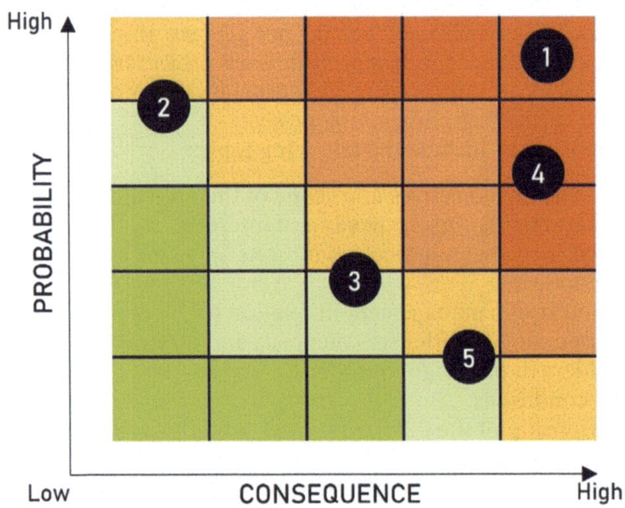

Figure 16 Ranking risks in a probability-consequence matrix. The risk factor is usually reduced by the relevant actions – be it preventive or 'plan B' identified by the risk analysis. Therefore, the risk moves from one position to another in the probability-consequence matrix as shown by arrows.

activities for each of the selected risks – both preventive actions and 'plan B' actions.

By identifying and discussing the reasons why the risk may occur, a better understanding of the risk in question is obtained. This makes it easier to specify the activities to be undertaken to eliminate the cause of the risk or simply reduce it.

Step 3: Risk management plan

A risk management plan contains two types of activities: (1) preventive actions that reduce the likelihood and/or consequences of unforeseen events and (2) a 'plan B' for each risk, mitigating the consequences in cases where the preventive actions failed to avert the adverse event.

In many cases, the project manager must accept some risks without fully robust preventive actions. There are incidents that cannot be prevented, for example, the weather can only be addressed to a certain extent. For other risks, preventive activities are not implemented because the likelihood and/or impact is judged to be small. In other cases, it is too costly to do anything about a risk. Some risks are so great that if they materialise, they could bring down the project or cause major damage to the project. In such cases, it is wise to eliminate the risk, for example, by changing the project objectives and/or the scope of the project. In the meantime, it is a good idea to change the project's approach or work with mock-ups/prototypes to test the difficult part on a smaller scale. In the latter case, the risk is continuously monitored until it is no longer relevant.

Some construction projects transfer risks to external or third parties. Risks are delegated through contracts, which clarify which parties assume the risks. Clients often shift substantial amounts of risks to their suppliers, but this may not always be economically sensible as contractors must charge for assuming the risk. Clients should analyse which risks are worth transferring and which are economically advantageous to keep in-house. Alternatively, risks can be transferred to insurance companies, which assume financial obligations in the event of the risk materialising. A risk analysis should also define the responsibilities of the parties involved in the project, making it clear who is responsible for managing specific risks. A general 'rule of thumb' is to strive to have the party best able to handle (usually mitigate) a risk also have the responsibility.

Risk analysis usually results in several specific activities to be initiated with responsible parties and deadlines and activities/milestones in the project plan. The planned risk activities will change the risk picture, that is, the probability and/or consequence will be reduced. If the costs of some of the planned risk activities are too high in relation to the expected risk reduction, it should be considered whether the project should instead choose to accept the risk and not implement the activity.

The discussions are documented in a plan or a risk log, which can be set-up in Excel or similar; an example of a risk management plan is shown in Table 4.

Table 4 Example of a risk management plan.

Risk	Reason (In Words)	Consequence (In Words)	R = S × K	Preventive Measures/ Plan B	Responsibilities
Requirement for wastewater treatment is not fulfilled	Wrong choice of technology	Fine from authorities	2 × 4 = 8	Seek for experiences with technology in other countries Make a list of other initiatives which can be implemented	PM
Breakage of water pipe construction	Incorrect assembly	Occupational accident Delay	3 × 5 = 15	Systematic quality assurance Suppliers assemble instructions are reviewed	PM

ITERATIONS AND FLEXIBILITY

The five key project management tools and their implementation have now been presented. To achieve perfect clarity, it is usually necessary to alternate between the tools until the results are synchronised.

The simplest process is achieved by drafting objectives. Then perform the stakeholder analysis and then go back to adjust the objectives. Then a milestone plan, then back to the objectives (maybe there were a few deliverables that didn't quite fit in the context). Then comes project organisation and, last but not least, risk analysis, which may again give rise to adjustments to the milestone plan. In certain cases, it makes sense to start by looking at the project organisation first, in order to get the project team planning the project quickly.

Through this work, both the project manager and the team will gain a deep understanding of the project. This will boost their inner motivation to start the project with full force.

It is important to understand that projects are inherently dynamic. This means that the target may change a bit at the beginning and therefore needs to be adjusted. Perhaps it is the preconditions that change. Even when the overall goals are fixed, the sub-goals may change, and hence there is a need to continuously integrate them into the planning.

It is better to bring the schedule 'into the common room', in the sense that the project manager should make it visible, acceptable, ambitious, and engaging. This is better than keeping it to oneself. When the schedule is developed jointly, everyone will show much more respect for the overall plan. And individuals can see themselves in the bigger picture: what is my contribution to the process?

And what am I doing to align with others doing their work – and not become the bottle neck? This is a better approach than planning everything from start to end in Microsoft Project by yourself and thinking that everything will go according to plan. Adjusting the plan dynamically may not feel like an optimal situation, but it's better to be realistic rather than stubbornly sticking to an early plan – regardless of how the project turns out.

Projects are almost always part of larger portfolios. In a way, your project is part of at least three different entities: one for the client, one for the consultant, and one for the contractor. The project may therefore be prioritised differently by the parties involved. For example, the project may be very important for one of the parties, whereas for the others it is just one of many delivery projects to be completed as quickly as possible and at the lowest possible cost. Business models, corporate strategy, and portfolio management can consciously or unconsciously influence each project. Therefore, communication and setting expectations is always important throughout the project.

Projects in the water industry always involve many parties with very different processes, cultures, competences, and organisations. As a project manager, you often get new 'employees' in each project, and only rarely do you have formal authority over your project participants. This makes it even more important that you with your manager put together the best team for the project and that you actively manage them daily. It is important that key people in the project are familiar with the overall framework and the main processes that influence the outcome of the project. In other words, we need to make sure that all parties across organisations and contracts recognise a clear common framework and common processes in order to be able to act smoothly when the project changes.

MAIN POINTS

It is the project manager's responsibility to ensure that the work with the five essential tools is carried out in a good process in the team. It is the project manager's task to facilitate these processes and ensure that the necessary iterations are made so that the result of applying the tools is clear and coherent. In addition, it is the project manager's task to ensure that everyone works within the jointly decided framework and in accordance with the decisions made during the planning process. Everyone needs to be based on the same foundation. This foundation is created by the group by using the five essential tools.

The five tools are:

(1) The project objectives should be clear and firmly adhered to by all project team members. The objectives are used to specify milestones in the joint milestone plan. The objectives and the milestone plan are the foundation for precise communication about the why, what, and where of the project. Missing or inadequate objectives increase the risk of the project running astray significantly.

(2) Stakeholder satisfaction is key in determining whether the project will be considered successful at the end. In particular, end-users are a type of stakeholders that need to be heavily involved from start to end for the life of the project. Their needs and interests must be incorporated into the project solutions, considering time, resources, and quality. But other key stakeholders also need to be identified and handled in a good way.

(3) The milestone plan is the overarching plan against which all other plans are organised. It is therefore important to start by identifying the main milestones in the plan. Larger projects have a number of different types of plans for different purposes, for example, short-term plans, task-oriented plans, risk mitigation plans, stakeholder engagement plans, and so on. They should all be synchronised with the milestone plan – which thereby becomes the 'master plan'.

(4) The organisation of the project needs to be considered carefully. There is a basic template for project organisations that includes a steering committee, a project manager, and a project team, as well as reference groups of key stakeholders. It is paramount to document the project organisation and make sure everyone understands their roles and responsibilities. Project organisation is not as straightforward as in an operational line organisation where everyone does the same thing year after year. The organisation may change as the project goes through various phases.

(5) Finally, get an overview and 'peace of mind' with a risk management plan. With it, you know that risks have been managed to the best of your ability. The project team will be able to avoid many adverse events, and if something does happen, the team will be prepared and can act aptly – thereby often being able to reduce the impact of the unintended adverse event.

The five tools must be applied on the project and be documented in writing at the start of the project and maintained on an ongoing basis. Specifically, the stakeholder and risk analysis should be updated regularly, and generally at least every 3–6 months. By being consistent with these five tools you will achieve a well-managed project. A well-managed project is a prerequisite for dealing with the complexity of sustainable challenges. 'Things take time' – so does good planning. Assign time for it – when looking at the lifetime of the project, it will be worth the time and effort.

YOUR REFLECTIONS

Grab your notebook and take some time to reflect.

Which tools or ideas in the tool descriptions are new to you and which elements do you recognise already?

Is there anything that surprised you, or is there an awareness that has been sharpened through your work with the project tools? Have you realised things you are not doing today that you should consider bringing into focus in the future? Do you have ideas for practising the new skills – practising tracks? Where can you get useful feedback?

In what way and to what extent have you used the project tools so far? The five essential tools are the foundation of good project management. How will you work with the tools? How do you organise the kick-off workshop or start-up workshop using the five tools – is there anything you can improve? Do you decided on your own principles for how to use the tools? When and how and with whom? Is there clarity in your organisation about how you best succeed in executing projects? In what way can this be improved? How can you be part of the improvement work?

Objectives

Think about your last or current project. Are the objectives laid out clearly and in writing? If they exist, are they communicated to all relevant parties? Are they easy to retrieve? Maybe you were not involved at all in formulating the objectives? How do you work with objectives in your organisation: clear or fuzzy? Certain or uncertain? Considering how central the objective is to a project, reflect on whether enough effort has been put into making it clear to everyone?

If you are embarking on a new project, carry out the objective-setting exercise with colleagues or already designated project participants. A session takes about 1–2 h, depending on the profile of the project, the experience of the participants, and your facilitation skills. Document your objectives in a PowerPoint or on a flipchart and consider how to communicate them to key stakeholders.

Stakeholder

- Whether you are a developer, consultant, or contractor, do you have a clear picture of each of the key stakeholders for the project as a whole?
- What do you do to keep the stakeholder picture updated in relation to all areas of the project – and how to do it specifically to your area of responsibility?
- How do you actively apply the stakeholder plan in the project?
- Some stakeholders are close to the project, and you meet them through dialogue meetings and so on. Have you deeply understood their perspective on the project?
- What are you doing to respond to stakeholders' needs, interests, and intentions?

Milestone plans

- How do you create the milestone plan in the project?
- Did you prepare an MS Project plan, or did you make a milestone plan together in the project team first?
- Why did you do it the way you did? What are the advantages and disadvantages?
- Have you used the overall objectives in your milestone plan?
- How do you make the milestone plan easily accessible and understandable for all participants in the project?

- High-level milestone plans may serve as a good communication tool (mapped out in PowerPoint or a similar tool). Do you actively use a simplified milestone plan when communicating with key stakeholders?
- How do you use the milestone plan to communicate with your steering committee?
- Do all project participants know the milestone plan and is it clear who is responsible for the different areas of action or milestones?

Project organisation

Before you continue, think about where you are right now – look around and place yourself in the current organisational chart of the project. Consider the following questions:

- Is the project organisation drawn up and easily accessible? Is everyone familiar with the description?
- Are the contractual arrangements reflected in the project organisation?
- Is the project organisation intelligent? Are clear and flexible workflows implemented? What can be improved? Can anything be made more flexible?
- Have you spent time going through the different roles and responsibilities? Does everyone know their own role and responsibilities? Do you know the roles and responsibilities of those you depend on to carry out your work?
- Is the role and responsibilities of the steering committee clear to all? Is it clear to the steering committee itself? Is it clear what kind of decisions the steering committee must take along the way?
- Does the project team have the necessary skills and competences? Are they available? What is the value of the project? Do they need training? Is external help needed?
- Which stakeholders should be addressed via reference groups?
- If the project is to be made 10% better, what competences and resources should be added to the project?

Reflections about risk management

- Is the risk management plan known to everyone in the project?
- Is there a clear delineation of risks and who is responsible for managing them?
- Is your manager familiar with the risk plan? Is the steering committee set-up?
- Is it clear how you follow-up on the status of the preventive actions?
- Is plan B up-to-date with the critical risks in the risk picture?

READ MORE

'Power in Projects, Programs and Portfolios' by John Ryding Olsson, Mette Lindegaard Attrup, and Niels Ahrengot (Olsson *et al.*, 2019) describes the basic elements, tools, and experiences in project management, projects, and portfolio management.

On the PMI website: https://www.pmi.org/ (PMI, 2021), you can find information on certification, membership, standards, and publications on project work, training courses, and so on.

The Project Management Body of Knowledge (PMBOK), the Project Management Institute's (PMI®) project management standard – is a handbook that describes PMI's approach to project work in terms of processes and knowledge areas (PMI, 2021).

https://ipma.world/ (International Project Management Association). On IPMA's website, you can find information about certifications, knowledge, training, and so on.

Individual competence baseline for project management (www.IPMA.world). The tool is used either as a self-assessment to target competence development or in connection with preparations for IPMA certification. The book can be purchased at IPMA.world.

You can read more about Scrum (SCRUM, 2023) at www.scrum.org and learn more about agile project management at: https://ipma.world/?s=agile+project+management or in IPMA reference guide ICB4 in an agile world.

You can read more about Lean construction at https://leanconstruction.org/lean-topics/lean-construction (LCI, 2023).

doi: 10.2166/9781789063349_0095

Chapter 3

Project manager of the future

Nobody made a greater mistake than he who did nothing because he could do only a little.

Edmund Burke

To succeed in sustainable projects, we must be more ambitious than our current approach. Today's complex projects often face numerous challenges, resulting in a stressful and chaotic environment. This, in turn, causes budget overruns that prompt cost-cutting measures, which often compromise sustainability goals and lead to exhaustion and discouragement. It is often the most ambitious project managers who bear the major brunt of this exhaustion. Regrettably, most people have yet to achieve a sustainable work–life balance in project management roles. To make progress, we must systematically address how to successfully execute projects to help the world transitioning to sustainability, while avoiding falling into old habits and conventional methods.

Transitioning to a professional, sustainability-focused project management approach presents new challenges for us as individuals. To achieve success in sustainable self-management, relational sustainability, and planetary sustainability, we must alter our core work principles. This requires changing ourselves and our approach to project management. It's not easy to abandon long-standing methods and habits, even when we recognise the negative consequences of maintaining the status quo and the positive possibilities of adopting new strategies. Clear objectives and a disciplined approach are essential until new practices become an integral part of our work routines. The transformation at the heart of this evolution will require conscious effort.

A common saying, often accompanied by a sigh of resignation, is: 'well, you can't change other people'. Although this statement holds some truth, it's also important to remember that we possess the ability to change ourselves. Gaining practical experience in self-development enables us to create an environment where others can evolve as well. In many cases, project participants unconsciously adapt their work style in response to a project manager's changes.

Authors: Pernille Ingildsen and Mette Thorkilsen

As a result, the onus falls upon the project manager to prioritise conscious and disciplined efforts in driving progress. Embracing change and leading by example are essential steps in this journey – 'you must walk the talk'.

When attempting personal growth, some methods prove effective whereas others do not. For instance, engaging in an educational programme fosters change by cultivating our understanding of a subject, such as engineering, geology, or psychology. This process refines our focus in a specific area and inherently diminishes it in less relevant directions.

Throughout our lives, we have been taught to value discipline in education and work. However, much of this discipline is performance-driven, focused on external demands and requirements. Our education system often lays out a specific path for us, offering limited guidance on inner growth and self-discipline. For instance, we know that eating less and exercising more can help us lose weight, but how can we convince our minds, heart, and body to follow through?

Similarly, when trying to change our project management behaviour to support a sustainable society, we encounter the same challenges. How can we change ourselves when our environment doesn't provide a supportive framework for the transformation? Success requires innovative experimentation, self-awareness, and self-discipline.

The pressure on the project manager in a transformation process, and the need to be able to create progress on the projects simultaneously, places significantly increased demands on the project manager compared to just 10 years ago. It has crept up on us in a way that we have hardly noticed it – except that we are exhausted and sometimes can barely see our way through it all.

This chapter focuses on sustainable transformation and how you, as a project manager, can cope and evolve within this agenda. We explore the current phase of transformation (hint: we are at a midpoint), the future role of project managers, and how we can measure individual and organisational development. Furthermore, we look at two models that clarify the change we are in the midst of. One is the 'inner development goals' which is an international attempt to create a counterpart to the sustainable development goals (SDGs). These goals were born from the realisation that inner change precedes the outer change we seek.

Furthermore, we examine the 'leadership heart triangle' model, which describes the three main elements of the work that we must work to transform: work on self, work on relationships, and work on the greater good. The two models arrive at the same conclusion from two different routes. Namely, we must take personal responsibility through holistic understanding, inner development, and cooperation. This is the path towards sustainable project management and hence a sustainable future.

THE SUSTAINABLE TRANSFORMATION

In spring 2021, Danish Professor Steen Hildebrandt proposed a curve to understand the times we live in as shown in Figure 17.

The curve describes how during coronavirus disease-2019 (COVID-19), we reached a kind of 'tipping point' between two paradigms. The two

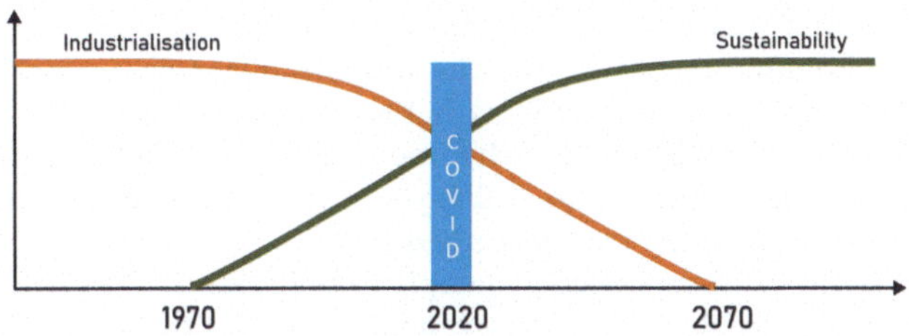

Figure 17 Industrial Revolution of 1760–1820 has created an industrial society to which we have become accustomed as normal. But a major change is coming, and we are probably living in the midst of what will later be called the 'sustainable revolution'. This drawing, which is a rendering of an idea presented by Steen Hildebrandt, places the COVID-19 crisis in 2019–2021 as the point where the paths of the two paradigms intersect – one on the path of development, the other on the path of dismantling.

understandings, the 'industrialisation paradigm' and the 'sustainability paradigm', reached a kind of temporary stalemate. Although sustainability is on the way as a basis for our organisation, industrialisation is slowly on the way out. Karen Blincoe, Ph.D., and author, calls this the 'COVID effect' in her book Achieving Sustainability: The Ultimate Human Challenge (Blincoe, 2022).

Some countries and sectors are further ahead than others in this transformation. But the transformation has become much more evident since the COVID-19 crisis. The multiple weather crises during the El Niño conditions of 2023 all around the world will further increase this focus as we get a foretaste of crises following the future normal.

Discussing sustainability is becoming increasingly prevalent and is generally less controversial in many contexts (though this is changing forth and back). Disagreements may still arise regarding individuals' assessments of the severity of the situation and its prospects, shared responsibility, or funding of preventive measures. In the big picture, however, a common understanding of the scale of the problem is emerging in most places – not as a science fiction, but as a current fact.

Still the mindset associated with industrialisation has not yet evaporated yet. Industrialisation still characterises how we think and make things: our culture, language, decisions, actions, products, monetary system, and so on. In our choices of which key elements underlie appropriate decision making we still see a lot of conservativisms. It is apparent that our actions, systems, and words have yet to catch up with our understanding of the sustainability goals.

The transition to sustainability has a major impact on the life of a project manager because:

(1) The basic premise of projects has evolved, requiring you to solve the usual tasks while also pioneering the creation of a sustainable society through the resulting project deliverables. Currently, we are collaboratively exploring practical ways to achieve sustainable water systems. As a project manager, it is your responsibility to actively participate in this innovative and experimental work.

(2) This means that you need to effectively work with and navigate both the sustainable narrative and the industrial narrative simultaneously. This often results in dilemmas between the two narratives. Whether you resolve the dilemmas to the one side or the other or with a nod to both sides depends on your skills, your creativity, and the depth of your understanding of the two narratives.
(3) Your results will be judged by people in both narratives – and it will not always be possible to satisfy everyone. Thus, it's crucial to enhance your capacity for finding smart compromises and ways to include both while also strengthening your resolve to stand firm on the side of the future. To do this, it requires a deep understanding of the sustainable age we are entering.
(4) The complexity of projects has been on the rise for quite some time. In this age of transformation, complexity continues to grow. Although industrial society standards remain largely intact, altering production chains proves challenging. The industry's focus on efficiency and profits generates resistance to new ways of thinking and working. Mapping product life cycles and supply chains as well as the practical construction and assembly adds to the complexity. It can all seem rather messy, to say the least, both for you, as a project manager, and for all of those affected by the project. It is your job to try to communicate clarity in structure and overview.
(5) Stress and burnout are nowadays not only a remote possibility, but also a near-term risk that you must deal with almost on a daily basis; not only for yourself, but also for project participants, advisers, suppliers, steering committees, and stakeholders. Although you cannot take responsibility for everyone, you must work to create a supportive and collaborative environment where individuals can thrive without compromising their mental or physical well-being.

Project management has always been complex and tumultuous, and it shows no signs of becoming simpler. Despite these challenges, there is great potential to make a positive impact. Projects that create tangible and physical changes in the world drive the green transition and replace outdated structures. The stakes are high. Consider it a privilege to contribute to these transformative projects; strive to deliver impressive and effective results that you can be proud of.

Project management has long been the slightly messy cousin in the industrial society family, not quite fitting into the predictability of the assembly line and the spreadsheets of finance departments. Right now, we need to take project management even more seriously than ever – as project managers, we need to take ourselves and our discipline seriously. We need to find clear and communicable ways to make sustainable projects the centre of transformation for the next decades. The future will be created through projects.

It is literally through projects that the changes towards a more sustainable society will manifest themselves. For water projects, it will manifest itself in our shared physical spaces – both public and private, as well as the infrastructure that surrounds us and binds it all together and makes it work. The industry creates the basis on which the sustainable societies and people of the future will thrive. Because the transition to a sustainable society will be driven by changes implemented in projects, project management is a key success parameter in all organisations.

The ongoing transformation necessitates the ability to discuss and justify decisions in two distinct worldviews: industrialisation and sustainability. It is necessary for the project manager to have a language for both. Practising speaking concretely into both stories creates an understanding of the situation in both systems and a capacity to find the smart solutions that work in both mindsets. However, as project managers, we must also insist on placing our attention in a vision of the future of society. This is why innovation is a recurring theme in companies and organisations today. Innovation is a technical response to the needs and challenges we face both now and in the sustainable society of the future. We are building in the society we live in now, while at the same time trying to understand what the new society should contain and how it should function in a future that is likely to be very different from what we know today.

Being a project manager during this transition between two distinct mindsets presents numerous challenges. One of the challenges is that there is a constant conflictual dialogue between the two mindsets. The conflicts can be seen between those who have their feet planted deep in the industrial machinery and for those seeking to establish themselves in sustainable greenery. But it also subtly plays out within us as we, natives of the industrial paradigm of cars, iPhones, and central heating, try to figure out how to navigate the new sustainable universe of waste sorting, more space for nature, and shortening of work hours in order to flourish. This requires a form of long-term view combined with a high degree of flexibility and agility.

It is a transformation that is particularly complicated now because the two mindsets exist in a way that they are jumbled together, and it is difficult to always distinguish which is which. This means that success criteria easily become contradictory and unspoken expectations create conflicts before, during, and after project implementation. There is therefore a continuous need to negotiate between the two mindsets. The developmental stage of the industrial paradigm combined with the new arrival of the sustainability paradigm adds layer upon layer of complexity on top of the project management processes: benchmarking, digitalisation, innovation, sustainable construction, environmental impact mapping, life-cycle analysis, total cost of ownership, personal development, partnering, tight financial management, quality standards, occupational health and safety, stakeholder involvement, and so on.

All these good and sensible focus areas are competing for attention and focus. It also creates an increasing number of actors and specialisations. Altogether, it requires a different type of project management, where cooperation, trust, and relational competence become a central focal point.

ACT ON BEHALF OF SUSTAINABILITY

Project management must be understood as an independent and specific discipline – often carried out in conjunction with a technical discipline. You might be tempted to think, 'Well, well, well, we've been dealing with project management for years, so there's nothing new about it'. But the real question is, have we truly been practising it for years? Have we approached it with the required level of seriousness?

There is a wide range of theories, concepts, and systems for project management, but putting them into practice with the seriousness of an Olympic discus thrower, pole-vaulter, or archer – we are not there yet in the industry.

When we witnessed China build an emergency hospital for 1000 patients in 10 days in 2020, it became clear that the bar had risen. The hospital was completed on 3rd February and received the first patients on 5th February. Not because we believe that 'project management' is a competition, but to point out that for most project managers there is still potential to improve. It is crucial that we acknowledge and strive to unlock this potential to successfully embark on the sustainable transition.

Poor and ineffective project management has a negative impact on finances, green initiatives, and working relationships. It takes a toll on individuals when one project after other leads to a stressful and exhausting work life. There is a growing desire among people to actively contribute to the green transition through sustainable solutions. However, it often becomes challenging once projects are underway. Financial pressures often force projects to compromise on innovative and sustainable elements. It is crucial that we improve our ability to successfully realise visionary and sustainable projects.

Project management is a comprehensive discipline that encompasses a range of management tasks. These tasks involve setting objectives, planning milestones and activities, organising and motivating teams, monitoring and ensuring progress, managing finances and budgets, promoting customer satisfaction and stakeholder engagement, reporting on project progress, addressing conflicts and making necessary corrections, managing project changes, handling risks and dilemmas, ensuring optimal quality, and closing projects while promoting learning.

A week in the life of a project manager is often very mixed and can include team meetings, working on schedules, talking to a team member about his task, briefing the steering committee chairman on problems that have arisen, reporting the month's progress, comparing the realised spending with the budget, visiting the site and discussing health and safety with the chairman, attending construction meetings, and discussing different solutions to a technical problem that has just arisen.

The project manager's development path

The project manager plays a vital role in determining the success or failure of a project.

The difference between a project led by a skilled project manager and an inexperienced project manager can hardly be underestimated.
Examples are as follows:

- Well-being and learning on the project: they make working on a project a pleasure for all participants because they provide structure, security, and momentum.

- Value for project owners: project owners feel safe when planning and execution are clearly communicated, and clear choices and recommendations are provided along the way.
- Ease of use for end-users: good project managers can integrate and respond to multiple needs in the project design and customise the project to make the end product more valuable to those, who will use it. It is often difficult to capitalise, but user-friendliness saves irritation for those who must work in and with the facilities resulting from the projects.
- Sustainability: addressing sustainability aspects also has a huge impact, which is difficult to measure, but which nonetheless goes into the large pool of environmental impact reduction.
- Finances: they provide a continuous overview of the finances and ensure that the project is sustainably within the agreed framework all the way to the target. They find the smart solutions and implement them so that the savings are realised.
- Innovation and rethinking: they create spaces in the project for rethinking and innovation and manage to materialise the innovation results. New thinking and innovation only have value if the project is well managed, otherwise they will not have the intended positive effects. New thinking and innovation are only possible where everyone feels safe in the projects.
- Risk profile: if risks are managed proactively, they do not take up much space in projects – alternatively, they can take up a lot of space and seriously delay and cost projects.
- Level of conflict: similarly, conflict levels are lower in well-managed projects. In the best case, disagreements do not evolve into conflict but are resolved effectively.

The project manager's approach, overview, and leadership influence the project in a myriad of ways. In this way, the value of a skilled project manager can easily be 5–10 times higher than an inept or immature project manager.

It is the project manager who has the broadest overview and the deepest insight into the facets of the project – without necessarily knowing all the details. It is not always the case that a skilled project manager can save a project. Events beyond the project manager's control can sometimes 'topple a project'. However, even in very difficult projects, the project manager is the key to avoid most of these difficult situations, to make it work anyway, or to fix a project that has gone off track. The project manager is the linchpin. Therefore, the 'project management maturity' of the project manager is of key importance and all organisations would do well to invest in the development of their project managers. Investing in education, training, mentoring, and coaching will almost always pay off when it comes to developing project managers. Good project managers are generally interested in and constantly seek out opportunities to improve – whether it is through courses, self-study, books, sparring, coaching, or advice.

There are many compelling reasons to pursue a career in project management and to take pride in excelling in this role. Some are attracted to the project management challenge. Part of the challenge is that there is no certainty of success, and the success or failure of the project depends to a large extent on the project manager.

The project is a holistic endeavour, and the responsibility of the project manager is to ensure its comprehensive nature and functionality. The project manager literally has his or her hand on the steering wheel and plenty of opportunity to make decisions himself or herself. They must also face the consequences of both good and bad decisions. For many people, navigating a project and making the decisions themselves is a great motivation. There is often an opportunity to make your own mark on the project, to include new technologies and in many other ways 'show what you are good at'. A project is almost by definition a challenge – or, as in the basic meaning of the word, a throw ('ject') forward ('pro').

Projects attract a certain type of people who may not thrive in a 'regular job'. They are typically people who not only thrive on change, but who also want to be real change agents. They are constantly looking for inspiration and new ideas. Project managers are often both generalists and specialists, able to jump freely between the general and specialised approach to the work. Without specialisation it doesn't work, without generalisation it doesn't work. They are people for whom the world is exciting and who are constantly looking to learn new things. For the project manager, learning is always motivated by putting the new knowledge into practice, and what better place to do that than in projects. They are curious people and often ambitious; ready and confident to 'create the future'. They see the big picture and bring it all into play to create the future. For those who thrive on project challenges, it is difficult to find another endeavour that offers the same level of excitement and fulfilment.

For some individuals, project management serves as a stepping stone in their professional growth, providing ample opportunities for development before transitioning into a line management role. Project management provides a microcosm, where it is possible to experiment with what works and what doesn't in almost all leadership disciplines. We would always choose a former project manager as a line manager – regardless of whether the project succeeded or not. Project management creates an opportunity for immediate and direct feedback on management style and decisions; feedbacks that are slower and less visible in line management. By working with project management, one can learn which management principles are effective and how to apply them in practice.

For others, mastering a subject can be interesting until it becomes too easy and feels like repetition. This is where it can be a good idea to step up and deal with projects at a systemically higher level.

In any case, project management is an opportunity for professional and personal development and there will always be opportunities to fine-tune your management skills.

Inner transformation

There is a need for the project manager to be able to manage the increasing complexity of projects. Sustainability requirements add extra pressure on projects, often complicated by new methods, considerations, decision-making paths, and arguments. Sustainability also necessitates working with a responsibility that has a larger and more abstract scope than previously. The complexity will often exceed the capacity of a single person, so there is an even greater need to be proficient in collaboration and, not least, to create and trust each other. If the project manager and project organisation have not reached a relatively high level of development, it makes the transformation even more difficult.

Nobody profits if the project manager buckles under the pressure of the sustainability imperative. Therefore, while recognising the potentially serious implications of the sustainability challenge, we also need to be in a way 'lighten up'. We need to adopt an inclusive and humorous approach. We need to establish personal practices that make us flexible and prepare us for failure – both on our own behalf and on behalf of sustainability. Not everything is possible – even though we often know what the right thing to do is, it is not always possible. The project manager can easily end up having to take the lead on quite difficult agendas and messages.

It's easy to think that big changes mean that everything must be done differently. But 'don't throw the baby out with the bathwater'. The five essential project management tools presented in the previous chapter are still central to every professional project manager – and perhaps even more crucial than ever before. Managing and succeeding with complex projects without the five essential tools is or will become nearly impossible. So be disciplined and keep using the tools. But – and this is a big 'but' – the tools won't help if we don't focus on creating an 'inner workspace' that is clear, balanced, reflective, and both visionary and grounded.

STAGES OF DEVELOPMENT OF PROJECT COMPETENCES

How can you measure whether you are moving in the right direction? How do you know where you are on your developmental journey? It is often difficult to measure what really counts. Einstein has quoted: 'Not everything that can be counted counts, and not everything that counts can be counted'. Measuring progress is difficult. There are several step-by-step models that indicate the level of maturity of project management. International Project Management Association (IPMA) is a well-developed system for evaluating a project manager's project management competences. Overall, certification is available at levels D (project associate), C (project manager), B (senior project manager), and A (project director, programme director, or portfolio director). A certification at IPMA is based on the evaluation of a set of specific competences under the three main areas: people, practice, and perspective. Competences include negotiation, design, teamwork, communicating effectively, and strategy. The IPMA system constitutes a good and thorough evaluation system of the many competences acquired over time. In addition to an evaluation and certification, the IPMA system is suitable to ensure a reflection on one's own project management.

There are several alternative project management maturity models available on the Internet and in the literature. They typically have five or more stages

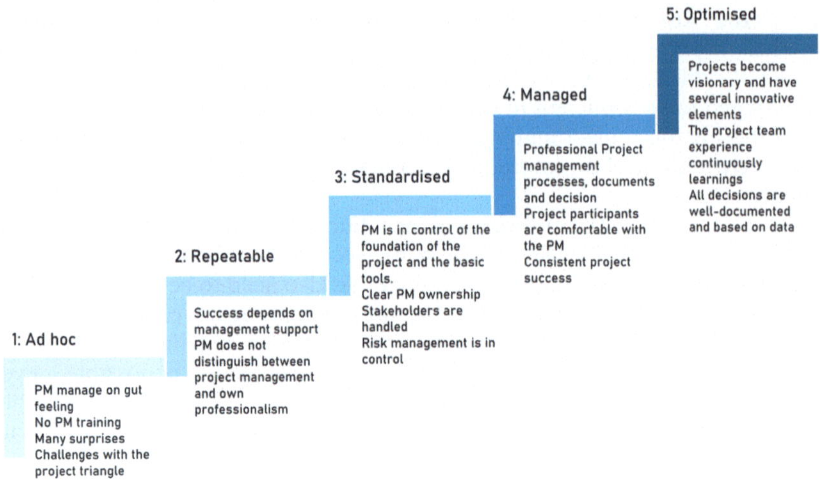

Figure 18 Development stages of project manager skills.

with an explanation of the characteristics at different levels. We have not found one that we think is spot on but have instead chosen to be inspired by several different ones. The model shown in Figure 18 is an example and a suggestion, but there are many ways to specialise and improve your project management skills – and there are many different development paths to choose from.

Maturity of project organisations

Project management does not exist in a vacuum, but also depends on the organisation you are part of. It is generally easier to do good project management in an organisation with a high degree of project management maturity and much more difficult and resource-intensive in organisations that are either new or immature in their framework and experience with project management.

The maturity of organisations is about a wide range of different issues:

- Management of the project department
- Project management culture
- The other project managers (a skilled few can raise the level of all)
- Availability of relevant and good tools
- The experience of the surrounding departments in interacting with and servicing projects and programmes of project leaders

The organisation, through its culture, governance, and management style, often effectively sets both the ceiling and the floor for the quality of project management.

Similar to the maturity models for individual project managers, it is possible to estimate the project maturity level of an organisation and its aims, the model below is only an example. In this way, the model also becomes a way to clarify the organisation how and with which steps the organisation should

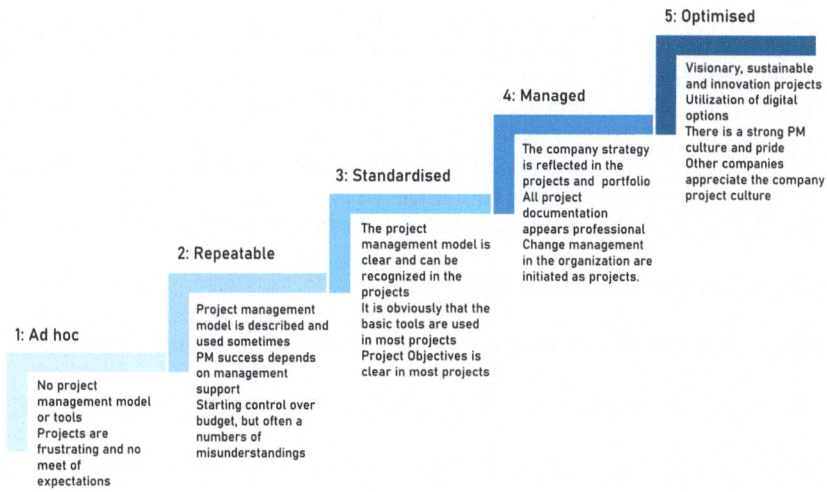

Figure 19 Development stage of an organisation in terms of its overall ability to provide a framework for good project management.

develop and actually do develop. You can have a meaningful dialogue about the necessary criteria to progress through different steps. This discussion will result in effective learning and the establishment of a well-defined strategy for organisational development. Through this process, the development path becomes clearer and enables the setting of specific goals (see Figure 19).

THE INNER DEVELOPMENT GOALS

In 2022, the 'Inner Development Goals' (2022) were developed in response to the SDGs. The non-profit organisation Inner Development Goals (IDG) is a coalition of people who acknowledge the insufficient progress achieved between 2015 and the present (2022) in comparison to the targeted achievements by 2030. Hence, they recognise that something different is needed. IDG works according to the principles of open source and has initiated their work based on modern research that shows that the inner abilities we have as humans and will need towards 2030 to succeed must be developed.

IDG's intention is to accelerate the work on our human resources to better match the ambitions of the United Nations' (UN) 2030 SDGs. The existing complexity around the SDGs must be supported by the right competences and skills, as well as an inner development for the individual. We share this insight and believe that this is a key focal point if we want to succeed in adding the word 'sustainable' to the title 'project manager'.

In the climate and sustainability world, there is simply too much talk and too little action. Now is the time to act, to start our own inner development process if we are to fulfil the UN SDGs. Technology will help us part of the way, but our lack of inner competences and collaborative relationships will make the road unnecessarily long compared to the time we have left. It is time to turn talk into action. We need to act in relation to ourselves, our relationships, and the industry. This translation into action requires inner development.

The IDG model was launched in 2022 and has initiated an important dialogue that the time has come for us as humans to actively engage in our own inner development. The IDG organisation identifies five overarching themes with a total of 23 supporting outcomes (Inner Development Goals, 2022); use the goals as inspiration in your daily work to become a sustainable project manager

(1) Being: attachment to oneself

Cultivating our inner life and developing and deepening our relationship with our thoughts, feelings, and body help us to be present, intentional, and non-reactive when faced with complexity.

Skills and competences

- Inner compass
- Integrity and authenticity
- Openness and learning mindset
- Self-awareness
- Presence

(2) Thinking: cognitive skills

To develop our cognitive skills by taking different perspectives, evaluating information, and understanding that the world must be seen as a coherent whole is essential for wise decision making.

Skills and competences

- Critical thinking
- Complexity awareness
- Perspective skills
- Sense-making
- Long-term orientation and visioning

(3) Relating: caring for others and the world

Valuing, caring for, and feeling connected to others, such as neighbours, future generations, or the biosphere, will help us create more just and sustainable systems and societies for all.

Skills and competences

- Appreciation
- Connectedness
- Humility
- Empathy and compassion

(4) Collaborating: social skills

To make progress on common concerns, we need to develop our ability to include and communicate with stakeholders with different values, skills, and competences.

Skills and competences

- Communication skills
- Co-creation skills
- Inclusive mindset and intercultural competence
- Trust
- Mobilisation skills

(5) Acting: enabling change

Qualities such as courage and optimism help us to achieve true freedom of action, break old patterns, generate original ideas, and act with tenacity in uncertain times.

Skills and competences

- Courage
- Creativity
- Optimism
- Perseverance

MANAGEMENT WITH THE HEART

Some project managers possess the skills to create extraordinary projects where both outcome and process work in a healthy way. This is the goal of all sustainable project managers. To achieve success is about embracing the leadership role and pursue a growth journey as both a leader and a human being. The journey is lifelong and provides continuous new insights and opportunities to achieve a deeper leadership practice. It is a task in the Kierkegaard sense: to become a sustainable project manager is to step into character. Keeping on the journey and keeping a learning mindset is a prerequisite for creating the change that is needed. We must find a way to actively engage in our projects, overcoming challenges, and creating projects where not only the outcomes, but also the process of achieving them reflect a brighter and better world compared to what we see today.

In their book 'Sustainable Leadership: Leadership from the Heart', Sten Hildebrandt and Michael Stubberup describe a personal leadership model they call 'the leadership from the heart triangle' (Hildebrandt and Stubberup, 2016). We have returned to this model again and again because it fits well with the main themes described in this book. 'Leadership from the heart triangle' indicates three focal points for leadership understood from the perspective of a living system:

- Integrity: based on an inner focus centred on one's own person and personality
- Resonance: Building relationships through effective communication with individuals and groups
- Shared responsibility: understanding oneself as part of a larger whole and committing to the collective processes within that whole

The management triangle is shown in Figure 20.

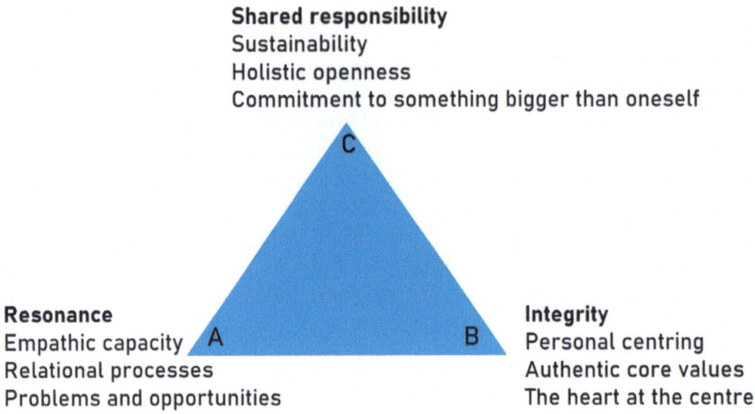

Figure 20 Leadership from the heart triangle is a leadership model based on the leader's capacity to functioning as a part of a larger system (Hildebrandt and Stubberup, 2016). The triangle has three areas of focus: integrity (a), resonance (b), and shared responsibility (c).

The three focus areas of integrity, resonance, and shared responsibility each have their own significance for the advancement to 'sustainable project manager'. There is, so to speak, a continuous focus and therefore also a continuous development process within each area; a development that requires concrete work and is characterised by the project manager's own practice.

The three processes are as follows:

- Integrity: a movement towards an individual inner practice
- Resonance: a movement towards empathic and collaborative relationships
- Shared responsibility: a move towards holistic understanding and responsibility

Integrity: a movement towards an individual inner practice

The shift towards personal integrity requires a lot of development work. In professional contexts, this type of development work has previously been reserved for high-level managers through direct coaching. However, personal development is increasingly becoming a prerequisite for good project management as well. Much development work starts with us.

IDG 1 (Inner Development Goals, 2022) is about deepening our relationship with our thoughts, feelings, and body; it speaks in the context of personal practice, which typically includes some form of mindfulness, meditation practice, yoga, or body-awareness sport. There are many possibilities. What the methods have in common is that they create an increased capacity for reflection and a reduction of our automatic reaction patterns to external provocations. A project manager who rapidly and instinctively responds to disruptions might initially seem effective; however, the opposite is true. Reflection is crucial before taking action, as the gap between an incident and reaction offers the opportunity to act more wisely than before.

It requires a special type of skill to manage one's own inner landscape of thoughts, biases, emotions, shadow sides, and assumptions. This inner management is necessary to create and maintain an open, present, and learning approach to the world. As an added benefit, a strong understanding of and connection to oneself allows for increased adaptability and resilience when faced with anxiety, stress, and burnout. As you develop the ability to respond wisely to external cues, you also gain the capacity to respond wisely to your own internal cues – provided you train to be more conscious of these cues.

To act as a transforming agent for a sustainable world, a project leader must make the sustainable agenda present in his or her own life and way of living. Project leaders with integrity and authenticity are needed. For most, this requires a clear inner practice and clarification of values and priorities.

Resonance: moving empathic and cooperative relationships

> 'Context, context, context', is what Stine Lene Hansen from the DanChurchSocial always says when she stops someone who is about to judge someone else. It is so easy to get frustrated with other people's actions, choices, and behaviour. Even though we recognise this, it can be difficult to accept that what we perceive as 'other people's poor actions, choices, and behaviours' are often reasonable and justified from the perspective of the individual involved. Therefore, before passing judgement, it is prudent to understand the other person's context – essentially, to 'walk a mile in their shoes'. This demanding process requires genuine interest in one another as human beings, listening to each other's context, openly sharing our own context, and approaching the situation with open-mindedness and non-judgement. We must suppress our innate urge to judge and focus on a sincere effort to understand instead.

Specifically in a project context, this means changing the fundamental mood from a rights-based approach, using conflict tools and protecting ourselves against risks, to an approach based on mutual trust, cooperation, and empathy. It requires awareness and courage. It is very easy to fall back into old patterns and roles. There is therefore a need for the project manager to have the inner strength to stay focused on the relationship work when he/she is challenged by relationship issues. It is not always as easy as it sounds. Working to create stronger and more empathetic relationships is both an intellectual and an emotional endeavour. Cognitively, we need to be aware of the 'traps' we so easily fall into when we engage in relationships with each other. Concurrently, our emotional responses must be taken into account to address imbalances early on before they escalate into full-blown conflicts.

To help us do this work, new forms of cooperation are emerging in the industry. In recent years, relational contracts, such as partnering and strategic partnerships, are emerging to replace the classic transactional form of tendering and contracting based primarily on agreements on specifications, quality requirements, price, and compensation. Partnering and strategic partnerships rest on a couple of assumptions. Firstly, that it is possible to construct a contract so that the client–contractor relationship changes from a win–lose game or often a lose–lose game to a win–win game. A win–win cooperation ensures both parties benefit from a successful project. This requires agreement on the project's objectives and mutual benefits. Trust, openness, and cooperation are essential for creating this atmosphere. A natural fit between the personalities of those involved can foster these qualities, but relying solely on chance is an unstable foundation for collaboration. Thus, it's crucial for the project manager to be able to facilitate the relationship-building process, especially when personality conflicts arise – or have a professional facilitator manage this. It is essential that steps to building relationships are taken before the actual work begins – often through social events designed to bring people together to get to know each other, find common interests, and understand each other's contexts. The lack of investment in relationship building is like sowing seeds on barren ground.

Shared responsibility: moving towards holistic understanding and responsibility

The shift towards a more holistic understanding and responsibility signifies that the entire Earth and its processes become central to our perception of responsibility. This can be interpreted differently based on individual perspectives. However, viewing it through the lens of planetary boundaries is a helpful starting point. It suggests a significantly different set of principles for living compared to industrialisation. It involves the creation of an altered broader moral foundation for decision making. For many, transitioning from 'team myself' to 'team whole Earth' is a lengthy journey. This journey usually entails a series of intermediate stages with escalating levels of responsibility like 'team me and those I love', 'team my country', and 'team eight billion people'.

Some inner confusion is natural as grapple to understand what this new responsibility means: 'am I responsible for all eight billion people and the entire planet?' Resistance may arise, perhaps best summarised by the doubt 'but we can't save the whole world!' So, how do we land this shared responsibility firmly in our inner psychology? The development of an ethical compass that takes into account the overall impact of our decisions is crucial. We must all carry our 'few grains of sand' to the new 'sandbox'.

Sustainability requires a radically different way of thinking about actions and decisions, and hence, we must question many of the things that we have not thought about earlier. Actions based on automatic thought patterns must be questioned. We cannot change thought patterns overnight – as we can see very clearly in the slowness that characterises the transition process. However, the surrounding crises underscore the pressing need for change. There is plenty to accomplish. However, if the entire team of 8 billion commits to a joint future, we can achieve significant results and would potentially see rapid progress.

MAIN POINTS

The world is at the centre of a radical shift from the old industrial paradigm to a new sustainable one. This transition contains dual extreme results: the potential for a future where society and livelihoods collapse or a future that offers a much-improved sustainable society. Project management plays a crucial role in ensuring the success of the myriad of projects that is the content of this transformation.

The transformation creates several cross-pressures for projects, project managers, and project participants – indeed, in general, anyone working with projects. Here in the 2020s, where this book is written, there is a kind of middle ground between the two mindsets. This makes it necessary, in a way, to live in both the industrial and the sustainable world and understand both. This is the starting point for creatively creating the new sustainable future.

If, as a project manager or as an organisation with projects, you are not competent and have a reasonable level of maturity in relation to project management, it will not only be difficult to succeed with the projects in the classic sense, but also be difficult to succeed with the sustainability agenda. To achieve a sustainable transition, substantial investment in developing project managers and promoting a robust project management culture is required.

Preparing for and adapting to this altered responsibility demands a change in mindset and internal orientation is needed from project managers. Clarity regarding personal leadership goals and legacy is essential. Both intrinsic sustainability objectives and 'leading from the heart triangle' highlight similar pivot points – the transition involves more than physical engineering; it requires examination of our 'inner software'. The world is shaped by our perceptions, so we need new visions and dedication to actualise a sustainable future through these images. A significant portion of this effort comes from executing projects. Therefore, your inner self is of utmost importance as a project manager, prompting questions:

- What is your personal responsibility for the whole?
- How do you manage yourself in relation to that responsibility?
- How do you succeed in building strong relationships and facilitating cooperation on responsibilities?

YOUR REFLECTIONS

Do you recognise that the COVID pandemic served as a catalyst for more holistic thinking? Many actions, such as testing, mask-wearing, vaccinations, social distancing, and isolation were driven by the pursuit of a common good, protecting both the vulnerable and society at large. Do you acknowledge that this global crisis made us realise the extent of our interconnectedness – breathing the same air? Additionally, can you identify areas where our response could have been better, instances where well-intended efforts went awry, risk assessments misjudged the situation, solidarity was not present? How might our lack of knowledge have led to inadequate or excessive security measures? How would you have managed the COVID crisis in your country, and what steps would a skilled project manager have taken?

Attempt to reflect on the situation using the five essential tools:

Tool 1: How would you have formulated the objective – again broken down into purpose, success criteria, and deliverables?

Tool 2: Which stakeholders should you be concerned with and how would you address them? Consider Trump and the US approach, European Union research and vaccine authorisation procedures, anti-vaccination protests, nursing homes, schools and day-care centres, and restaurants.

Tool 3: How would you define the milestones? What factors would prompt a change in strategy? How do uncertainties in mathematical models and estimates influence the right actions? What strategies would you employ to ease or strengthen restrictions?

Tool 4: How would you organise your team to deal with the crisis? Who would be included and who would be excluded? Which experts would you rely on? How would you have ensured the participatory (democratic) processes under time pressure?

Tool 5: How would you manage risks? Would you invest more heavily in prevention? How would you balance the benefits of prevention against potential mental health challenges and dissatisfaction? To what extent would you have provided better aid to Italy at the beginning of the crisis? What are the long-term risk consequences of not showing solidarity with poorer countries? What risks do other countries' Coronavirus responses pose? As a central project leader, how would you manage your own risk of becoming seriously ill?

Consider the answers to the above questions both in the context you are in now, where you can look back with the competence of hindsight, but also imagine your answers in a situation where you are constantly behind terms of information.

Also try to reflect on:

- Which countries had the best strategy through your eyes? What does their choice say about their values?
- What can we learn from trying to solve such a complicated project both from a local perspective and a global perspective?
- What have we learnt that can be used in the sustainability transition?
- What have we learnt that we can use in our projects?
- How can you best translate global responsibility into the scale of your current project context?

Try to build your own model of project management maturity. What would you call the five steps and how would you define them? Which competences would be important? Where do you place yourself and what would it take to get to the next level? What stands in the way? What changes in the organisation you work in would be most helpful for your development and performance as a project manager?

Assess yourself against the five intrinsic value goal themes in the IDG framework, including their sub-themes. A straightforward method is to rate yourself on a scale from 0 to 10 for each sub-theme. Are there specific themes where you're already focusing and making progress? In the coming week or month, where do you plan to focus your efforts? What is your development plan? Which aspects do you rate low and why?

What is your understanding of sustainability? Can you advocate for it under light pressure? Heavy pressure? Attempt explaining the key challenges to a friend (perhaps an imaginary one), discuss where you're making a difference, and emphasise its importance. Sharing this information with your project team can also help clarify potential knowledge gaps.

What is a sustainable project manager to you? What are their key traits? How well do your actions and mindset align with those perceptions? What steps can you take in the right direction? What actions can you initiate today?

READ MORE

Here are some suggestions for further study.

At https://www.innerdevelopmentgoals.org/, you can find out more about the inner world goals.

'Sustainable Leadership: Leadership from the Heart' by Steen Hildebrandt and Michael Stubberup advocates a new sustainable leadership mindset and new skills and competences needed to co-create a sustainable world (Hildebrandt and Stubberup, 2016).

'Achieving Sustainability: The Ultimate Human Challenge – Critical Barriers and Future Perspectives' by Karen Blincoe (2022) provides a different perspective on what it takes to succeed including Danish case stories of success including the experiences of Samsø, Novozymes, and Brahma Kumaris.

Several books offer different project management maturity models, see for example: 'Project Management Maturity Model (PMMM): Proven Strategies and Practices for Achieving Project Management Excellence' by J. Kent Crawford (2021) or 'Project Management Maturity Model: Building Better Project Management Offices' by J. Kent Crawford and Susan Sutton (Crawford, 2021).

Individual competence baseline for project management (www.IPMA.world) (IPMA, 2022b). The tool is used either as self-assessment to target competence development or in connection with preparations for IPMA certification. The book can be purchased at IPMA.world (IPMA, 2022b).

Part 2

Learn to Lead Yourself

doi: 10.2166/9781789063349_0117

Chapter 4
Be true to yourself

Being yourself in a world that is constantly trying to turn you into something else is the greatest achievement.

Ralph Waldo Emerson

The better you know yourself, the easier it is to be authentic in your leadership. The less you know yourself, the more difficult it is, and the more you are governed by your reactions to your surroundings rather than being based on your inner convictions. That's why it's essential to 'get a sense of yourself' and delve deeper into life-centred questions such as: who am I, what do I want, and why? To be true to yourself, you need to spend time reflecting on these deeper questions. There are many ways to do this; here we will present four different approaches: working with ethics, Kierkegaard as inspiration, understanding bias, and reflecting on ideas from stoicism.

Ethics provides a systematic approach to clarifying one's position in relation to the many dilemmas of being a good person. Thinking about being a good person often looks simple but practically, life is full of dilemmas, and we may not always succeed as well as we hoped. This can be a source of indifference or self-criticism. But when we delve deeper into a more precise understanding of our individual ethics, we gain many benefits in terms of leadership. We become more transparent to ourselves about how we want to be. We can decide on ethical principles we aspire to live by, we can recognise ethical dilemmas as they unfold before our eyes, and we can obtain better tolerance with ourselves and others because we understand the difficulties of making the right choice in difficult situations. We can extend the empathy of our self-care to other people in our projects and in life in general. Working systematically with ethics we also learn that ethics is not 'set in stone', but rather is a discipline where we can improve and deepen.

The stoics have much to teach us about how to act wisely so that we can live a life where we actually succeed in making a difference. As project managers, we often want to do much more than we can practically do. Often, what we

Authors: Pernille Ingildsen and Mette Thorkilsen

try to do in projects doesn't succeed because of many kinds of 'headwinds'. Sometimes we make things more complicated than they are, and other times we fail to recognise where we could have made things easier. Stoicism has gained popularity recently all over the world. This is possibly because stoicism provides a full set of advice on dealing with the difficult situations, life throws at us through clear thinking and deliberate action. There are plenty of reasons to think that everything is difficult, complicated, and overwhelming, but stoicism shows us the way to cut through the complexity.

Thinking fast or slow – that's the dilemma of researcher Daniel Kahneman. His research findings explain where and why we are fooled – by ourselves! When we think fast, everything feels easy and 'natural', but is unfortunately often wrong. So, why don't we just think slowly, analytically, and thoroughly instead? Because it's hard work. But if we know the critical situations where we as humans tend to get it wrong, we have a chance to stop in time and think long and hard. It can be worth a lot both as a project manager and as a human being.

Kierkegaard tries to guide us on the deciding question of what is most important in our lives. It is a difficult question that deepens the more we manage to revisit it in a skilful way. A journey through Kierkegaard's existential questions can lead us from an unconscious attempt to do what is right in the eyes of others to being increasingly in touch with our inner knowing of what is vital and right by our own standards. This releases us from the notion of trying to fit in and mimic everybody else or some kind of imagined average. Instead, it helps us being super-very-much ourselves – in the best possible way.

With this chapter comes an invitation to get to know yourself so that you can be true to what is really you. The more grounded you are in your own selves, the better you will be able to handle the challenges of project management and establish the authentic leadership we are all longing to be a part of.

REFLECTIONS ON ETHICS

Ethics deals with the complex and central question of what it means to be a good person. It is often a relatively simple question in an unreflective day, but if you dig deeper and look more carefully, the world is full of dilemmas. Thus, having a clear understanding of one's values can assist in navigating ethical dilemmas in project management, ensuring that you don't feel stuck or make incorrect decisions when facing moral challenges.

Examples of questions for reflection:

- A problem has emerged in the project, resulting in a substantial additional expense. If someone on the team had identified the issue 3 months ago, it could have been avoided. However, no one considered asking the crucial question, and now the bill must be settled, with the steering group being informed. The responsibility can either be attributed to the designated individual in the specialised area where the problem happened or be taken upon yourself. You realise that you share partial responsibility because you did not update the risk analyses within the past 3–6 month interval. That would probably have revealed the issue. What do you do?

- One member of your steering group is particularly dominant and difficult to discuss project issues with. This individual pushes for a decision that, while simplifying the project, could bring significant negative repercussions for the organisation. Although you report to the steering group, you're uncertain about whether to inform higher-ups about your concerns. Doing so could protect the company from potential problems but may also be perceived as disloyalty to the steering group. What do you do?
- The level of conflict with a supplier is steadily increasing over time. You find the supplier lacking in pro-activeness and you suspect that they are not carrying out their work efficiently. The lack of efficiency leads to extra costs to your project. When you confront your supplier with your suspicion, they immediately deny that there is anything to it. You can now choose between taking the conflict further or eventually having to accept the extra bill. At the same time, you are pressed for time in your project and find it hard to devote the time and energy required to take the conflict. What do you do?

In these types of dilemmas, the options are laid out clearly requiring you to make a choice within a short timeframe. Other types of ethical dilemmas are more subtle and may easily go under your radar. You need to actively seek them out.

Examples of such dilemmas are:

- You are carrying out a project similar to projects you have done several times before. It is a relatively simple project, and you know how to complete it simply and efficiently. However, the world situation makes you aware of the problems of climate crises and biodiversity. If you put some additional energy and thought into the project, you could, for example, search for more environmentally friendly materials, specify the use of green electricity or hydrotreated vegetable oil biodiesel in the excavators, or take other measures that would increase the sustainability of the project. Each action would be something you would have to discuss with a potentially reluctant steering group, and afterwards, it would add complexity and risk to the project, which might make you look less successful. What do you do?
- Key elements of the project's success depend on you working very intensively for a period in the project. In particular, one of the consultant's professionals needs to put in extra effort, so that you can succeed in this aspect. However, you know that the person is under a lot of pressure from other projects, they are working on. It is a very conscientious consultant who is willing to help the project as best they can, but you also risk bringing out a severe stress disorder in him or her. What do you do?
- You are facing challenges at home, and your family needs you to be there to help solve the problems there. There are also severe problems in your project, and your team needs you to be present there to reestablish calm and ensure a clear plan to get the project to a better and safer place. You can't do both as well as the situation demands. What do you do?

The ethics also asks open questions, where you cannot choose between a few options but must choose an appropriate level:

- How much responsibility do you need to take for environmental sustainability in your project? How much effort do you need to put into it, and how should it be balanced against costs for your own company or the client you work for? How much of your personal resources do you need to devote to securing an environmental solution?
- What are your responsibilities for aspects of your project that you do not know about? For example, what is your responsibility if your project affects wildlife in an area, but no one focuses on it? If you don't know the greenhouse gas emissions and no one has asked you to investigate it?
- How do you weigh your responsibility if you are just a small part of a huge problem? Your contribution to reducing greenhouse gases is negligible in the grand scheme of things. So, does it matter? All individual projects have only a negligible share, yet the sum of projects is the problem.
- There are people in your project group who do not think that the water environment should be considered and have an approach along the lines: 'we do what we are told and no more discussion'. One particular person in the project has a different idea and sees the responsibility in the project as part of the legacy they are passing on to the world through the project – for them, the issue is paramount. The discussion continues meeting after meeting, delaying the project and making it difficult to move the project forward. How do you resolve the conflict?
- You recognise that your commitment and understanding of sustainability will put you in complex ethical dilemmas and conflicts with your close partners and colleagues. You realise that it is easiest not to say too much or know too much. How little knowledge is it ok to have? To what extent is it ok to be 'wilfully blind' or 'wilfully dumb'?

Ethics is not as one might imagine possible to reduce to a question of simply following a set of rules. Sometimes, the ethical choice may be to break the rules or to do something other than fulfil the wishes and expectations of the immediate supervisor. As described above, ethical issues rarely present themselves in neat, well-defined packages. Often, they are not simply dilemmas of choice between two extremes or along a scale. Instead, many issues in management – and in particular in sustainable project management – are multi-dimensional. That means there is a balance to be struck between matters of the planet's climate, your own well-being, your child's challenges, the company's general data protection regulation rules, bosses with inappropriate ego drives, your team's cohesiveness, and your own need for rest. Questions of choices, balances, and prioritisation are present all the time. What does it mean for you to act ethically? Can you defend your choice to those who know only half of the equation you are trying to balance? Can you look at yourself in the mirror at night? And ultimately – how will you think about your life on the final day?

Ethics is a creative practice; becoming aware of the ethical issues you constantly face can initially seem overwhelming. But there are often opportunities to address the issues proactively. On the one hand, this requires that you have the tools to deal with the problems – both with yourself and together in the project group. On the other, you must have some basic principles to guide you. Often the principles are relatively simple and obvious … until they meet the complexity of reality. Therefore, being lenient with oneself and one's ability to act ethically is also needed. Understanding the overall context is important, and reflection on the overall context is essential.

Our ethical sensitivity is expressed in virtually all our choices. As this sensibility develops, it may be encompassing a wider and wider range. Common human development illustrates this: as a child, one is most concerned with oneself and one's own needs, and this ego-driven ethics will always be present to some extent to ensure one's survival.

Gradually, one's loyalty to one's immediate group – family, circle of friends, school class, or workplace – increases. This group influences our internal reasoning for doing this or that. You stick out for your friends and don't snitch.

The group taken into account during ethical considerations eventually expands to include a more abstract group of people or a community. This next step can be exemplified feeling a sense of belonging to a nation, class, or profession. You want your nation, community, or profession to do well, and you feel a sense of belonging. Suddenly, your choices change.

The next step is to try to match our understanding of the world with a further extension of our care and loyalty to all of humanity. The inner understanding and idea that my actions affect people in countries far more exposed to the effects of climate change than I am – that I have an ethical responsibility that extends to the other side of the globe.

The next step is to extend this care further to all living beings – a kind of cross-species ethical sensitivity. This means that the well-being of domesticated and wild animals is also part of the worldview that guides your ethics.

The unifying theme of these expanding perspectives on ethics is that it is becoming increasingly difficult to grasp and not see the suffering or problems our actions cause others. With more significant and remarkable mindsets and horizons, ethical considerations can be said to come in different sizes: small, medium, large, and extra-large.

It is possible to broaden one's ethical wingspan through decisions that consider wider perspectives – a larger whole. Thereby, it is more ethically advanced to include your company's well-being in your choices than just looking at your own needs. Next step may be considering your business partners as well – or the entire industry. In this way, it can be said that the greater the range of horizon one can take into account in one's ethical considerations, the more developed one's sense of ethics and of being 'a good person'.

Another dimension in which to expand your ethics is to take ethical considerations into more aspects of your project and your life. That is, to extend ethical considerations from just applying them to a limited number of technical issues to also consider ethics in how you manage your project group, how you cooperate and act concerning your steering group, how you keep track of the project's carbon dioxide (CO_2) consumption, how and to what extent you involve users in the project, how you manage the risk of stress, the energy you put into the environment work, and so on.

Your personal life choices are also important to reflect upon. It's hard to keep the ethical flag flying high in projects while flying around the world on holiday, driving a gas-guzzler of a car, and loudly scolding your children. There needs to be a connection between your life and how you lead others. And even if you could imagine yourself acting completely different in one place than in another, it would be hard to keep the two mindsets separate and difficult to be authentic. Hence, ethics concerns your work as well as your private life.

Making ethical choices is both a competence and a question of 'ethical fitness'. How good are you at putting your ego needs aside when giving praise? Do you take the credit yourself for its success, or do you generously place the credit in the right place? It's easy in the heat of the moment to make choices below the level of ethics you're aiming for – and sometimes, it can be challenging to repair trust once you've taken the less ethical route on an issue. It can also happen accidentally, where you have overlooked aspects in your blind spots. There may be questions about what you should do yourself, what the group should do, how the project should move forward, the pre-conditions assumed on behalf of others, and administering the given mandate correctly.

A large part of the ethical dilemmas used in traditional ethics training is about choosing who to save or do good for. Often, benefiting as many people as possible is a natural choice; at least as long as the people in question have the same 'value'. The choice between saving 3 or 10 strangers is relatively simple: save the 10. But perhaps the balance changes if the 3 are your close relatives and the 10 are strangers?

Another example is that the 3 are doctors, and the 10 are violent criminals. There can also be other reasons that shift the balance; if there is a shortage of doctors in the local hospital, it may be better to save the 3 doctors than 10 children. This may seem counterintuitive, but on the contrary, it is also obvious that there is a difference between killing a mosquito and killing a cow. For various reasons, we find it easier to eat a pig than a dog, and so on. There is a kind of ethical value-difference at stake.

There are also ethical choices about what you can live with – to do or not to do.

There are ethics about being accountable for your actions both internally and externally. There is a large field of ethics about how we treat others – am I loyal, honest, generous? Can I be counted on? The golden rule, enshrined in all religions, is to treat others as you would like to be treated. An upgrade from there is to treat others as they would like to be treated.

There is also ethics related to systems. What do we do about the climate system, natural habitats, democracy, local communities, the legal system, the procurement system, and so on? Once you look, there are ethical questions everywhere you look. There is an ethical imperative in everything we do or don't do.

Our current knowledge level is larger than ever. We know and understand what many of the 8 billion people worldwide struggle with, are pressured by, or succeed with. We know the planetary boundaries and what it takes for 'spaceship Earth' to remain viable in the future. We know the laws of physics and the latest technological advances; we know about vast amounts of ideas, philosophies, and religions – and we have access to so much knowledge, most questions we may have, we can find the answer to almost immediately. This access to knowledge is an obligation. It is a crucial realisation that, on the one hand, we cannot solve all these huge problems alone, our contribution as a percentage of the overall solution is small. On the other, it is clear that everyone's contribution is indispensable. Therefore, it is also obvious that, on the one hand, we must put effort and dedication into finding solutions, but on the other, it is essential to realise that nothing is gained if we wear ourselves out in the process – we need to strive to solve our current problems and take care of ourselves in the process.

What is perhaps a little less obvious is that our inner work – although not as apparent to the outside world – is central to success. The oft-quoted quote from Albert Einstein that 'we cannot solve problems with the same way of thinking as the one that created it' hinges on this insight. Carol Sanford uses another metaphor for this central understanding (Sanford, 2022). She uses Boehm's theory of intricate and super intricate order to explain this apparent paradox that our inner work can be as important or more important than our outer work.

Sanford's point is that one of the best things we can do for the world is to develop ourselves and to create space for others – first and foremost those in our team, our steering group, our customers, and our stakeholders – to develop and fulfil their potential more and more. When we realise our full potential, we can succeed. Without it, we don't stand a chance. A substantial part of management thinking in projects must therefore be about internal capacity building.

Another difficulty is that no one knows how this transformation will play out. Some optimists see a wonderful green world in the future ahead of us, whereas some more pessimistically see a bleak future with 'crash-scenarios'. A useful way for us to work with our thoughts about the future might be to remember that we are still determining how the future will unfold. It is often challenging to interpret the surprising things happening all around us. Perhaps we can recognise that our task is not so much about predicting what will happen. Instead, the task is to engage positively with what is happening right now.

We must strive to achieve a balance among various ethical decisions – cultivating a nuanced understanding of what is truly feasible and effective. It is a balancing of many considerations and a dedication to finding the best possible solutions in the continuous stream of issues we encounter in projects. Ethics is not a hobby we can engage in during our free time only. Instead, it is part of a continuously lived life. That is yet another reason why setting aside time for reflection is so critical for all ours work.

PROJECT MANAGEMENT INSPIRED BY STOICISM

Stoicism offers a nearly 2000-year-old philosophy on living and developing into a good human being. The most well-known concept from stoicism is probably 'to take adversity with a stoic calm'. But the point of stoicism is not to be emotionless, but rather to instruct people to direct their attention, willpower, and resources in a disciplined way to where they can make a real difference and then translate the attention, willpower, and resources into action that makes a real difference. The focal point is constantly to be a good person by using your inherent wisdom in the current context – whether you live at the top or the bottom of the hierarchy. A recurring idea in stoicism is that regardless of context, you always have room to act right based on right thinking.

Stoicism was developed in the centuries 350 BCE to 200 CE, in a timespan of over 550 years(!). The three most famous figures embodying stoicism today include the late stoics, who built upon the foundations set by their predecessors. They were Seneca, who was an advisor to the emperor; Epictetus, who was born into slavery; and Marcus Aurelius, who ended up emperor of the entire Roman Empire after being adopted when his father died while he was still a child. The three stoic philosophers span a period of 200 years, from 0 to 200 CE.

Due to increased focus on personal development, there has been a renewed interest in stoicism in the West in recent years. These ancient texts can still inspire us 2000 years later in our everyday life, which is undoubtedly much different than when the philosophy was written down. Stoicism can be seen as a fundamental human philosophy, answering many basic human questions, and providing practical strategies for becoming a 'good person'.

Stoicism is embodied in a vast set of books and reflections. One way to work with stoicism is by reflecting on the multitude of concentrated stoic 'one-liner' quotations. This section on stoicism will take some of these quotes as a starting point and show how they can be actively applied to reflect on how to achieve sustainable project management.

This chapter is inspired by Ryan Holiday's work; he has written several books on stoicism (Hanselman, 2016) and describes a model for understanding stoicism through three basic layers: (1) rationale and discernment, (2) discipline of action, and (3) character and willpower.

Epictetus describes the three layers of stoicism as (Hanselman, 2016)

> There are three areas in which the person who would be wise and good must be trained. The first has to do with desires and aversions – that a person may never miss the mark in desires nor fall into what repels them. The second has to do with impulses to act and not to act – and more broadly, with duty – that a person may act deliberately for good reasons and not carelessly. The third has to do with freedom from deception and composure and the whole area of judgment, the assent our mind gives to its perceptions. Of these areas, the chief and most urgent is the first which has to do with the passions, for strong emotions arise only when we fail in our desires and aversions.

A central point for the stoics is to distinguish between what is within one's scope of power and what is outside of it, and then to focus on doing the right thing with what is within your power. Then it is a matter of consistently and without fear or hesitation turning thought into action. Many people tend to act only when it is almost too late – some not even then. Stoicism emphasises doing what you believe in and doing it directly and immediately.

Strength of character depends on willpower and direction. It is essential to set aside time for self-reflection and cultivate the stoic determination required for perseverance. A stoic follows through on their words, regardless of any challenges they may face.

A key concept is 'Amor fati', that is, love your destiny. So, if there is resistance to what you do after careful consideration, you are meant to meet resistance. You must then take this resistance without losing your temper or spirit. The clear results of this approach have made stoicism popular and widespread in high-performance settings such as in sports and the military.

Marcus Aurelius mentions the same three things differently (Hanselman, 2016):

> All you need are these: certainty of judgment in the present moment; action for the common good in the present moment; and an attitude of gratitude in the present moment for anything that comes your way.

Stoicism contains a treasure trove of quotes worth reflecting on. We have chosen two quotes within the three main areas to illustrate how you can use the quotes to reflect on your leadership practice. This may seem like a rather high-minded approach. But it turns out that the old stoics have an inspiring ability to repeatedly tell us what we need to hear and reflect upon. Stoicism has been a major inspiration for, among other things, cognitive behavioural therapy with psychology, where it is used to support self-management and resilience. Here, too, the ability to recognise what is beyond one's control and take action where one can is central. In this way, reflection on stoic quotes can help those who are stressed and anxious and those who are working to strengthen their self-management.

Two quotes on rationality and discernment

If we take a closer look at the first ability – rationality and discernment, the below quote stands out (Hanselman, 2016):

> The chief task in life is simply this: to identify and separate matters so that I can say clearly to myself which are externals not under my control, and which have to do with the choices I control. Where then do I look for good and evil? Not to uncontrollable externals, but within myself to the choices that are my own. Epictetus

This idea is at the centre of stoicism, to distinguish between what I can't control and what are within my own choices. The stoics believe that the only thing I can control is how I think and act. I cannot even control what the consequences of my actions are – at least not 100%. We all know situations where good intentions ended differently than intended. We can't even control what happens to our bodies. The body gets sick; it can be thrown in jail, and so on. But we can always choose which of the possible actions in front of us we take.

The stoics would tell the project manager that your task is to choose among the best options you can come up with using your imagination, talent, experience, advice from others, and so on. On any given day, you, as the project manager, choose how to apply your energy to the project. Is today the day to look at the milestone plan? Or is it today that you talk to the project participant who seems not sufficiently focused? Or is it today, that you study the CO_2 footprint of the various technology options before you? What do you choose today?

The quote is also relevant when conflicts are brewing. Instead of placing blame, we must look at our own role in the conflict and our options. Let's centre our attention on our best actions to help the dispute to get resolved.

Take the task of establishing and following up on a schedule as another example. It is within my control to ensure that the schedule is as good as possible and contains the best knowledge, imagination, and experience that my team and I can come up with. It is also within my control to get feedback from my project participants, suppliers, steering group, and other sparring partners. I can involve the right people and make sure that the process of establishing the schedule is well facilitated. But there is a point beyond which I can no longer manage the schedule – problems may arise, partners may not perform as promised, technologies might not work after all. Such problems are beyond your control. But then again, as these unintended consequences have happened, again the ball has been played into your courtyard: how do you respond to this new situation?

Following our top-of-mind impulse in a heated situation is rarely wise. Note that subsequent explanations such as 'I'm so impatient', 'I have a temperament', or 'I'm not into details', and so on are not worth much. With such explanations, the project manager introduces problems into the project through his unconscious and unexamined behaviour. This leads to the below quote (Hanselman, 2016):

> When you let your attention slide for a bit, don't think you will get back a grip on it whenever you wish – instead, bear in mind that because of today's mistake everything that follows will be necessarily worse. … Is it possible to be free from error? Not by any means, but it is possible to be a person always stretching to avoid error. For we must be content to at least escape a few mistakes by never letting our attention slide. Epictetus

No, of course, it is not possible to be free of errors in projects. Projects usually have so many details and details about details, that the hope of making it through without mistakes is hopeless. That's why it's a good idea to start by accepting that mistakes will happen.

But still, disciplined attention and focus are of enormous importance. Everybody on the team and you as a project manager, as a role model, must be focused while doing the work. You must use the best of your abilities and take a conscious and systematic approach to solving all the tasks at hand. It is unwise to work all night and turn up tired for the quality session the following day. Instead, the project manager is responsible for setting a good example with his/her ability to focus where it counts. In a sense you need to treat your ability to focus with respect.

Epictetus' quote tells us that when we turn up inattentive, we make more mistakes. And each mistake makes the project worse off. We can sometimes mistakenly have the opposite feeling because correcting a mistake may give us a good feeling of action and efficiency, while preventive actions might feel more tedious. The problem we tend to overlook is that the energy, creativity, time, and money spent on correcting the mistake are lost in favour of creating other types of value in the project. Because of the mistake and the lost attention, the overall project is worse off.

So, it's not a story about feeling guilty and shaming people who make mistakes. Instead, it's a call to stay focused and to be systematic in project management. From there on, we must let things play out and later to deal with the things that didn't work out despite all efforts.

Two quotes on the discipline of action (Hanselman, 2016):

> Don't you know life is like a military campaign? One must serve on watch, another in reconnaissance, another on the front line … So, it is for us – each person's life is a kind of battle, and a long and varied one too. You must keep watch like a soldier and do everything commanded … You have been stationed in a key post, not some lowly place, and not for a short time but for life. Epictetus

The stoics are not advocating fighting and aggression in this quote. Instead, the quote says something about the variety of jobs there are in life and, not least, that your efforts count. Project management – and life – is a complicated affair. You must have a strategy, be disciplined, and the same goes for your team. The project will not work out successfully without discipline and overview. Seneca is famous for the short quote: 'Vivere militare est' – to live is to fight. And projects are quite often 'a struggle'. Sometimes it seems like so many things are trying to bring the project down. In addition, if you want to make the project sustainable, that doesn't simplify the story the least. It's easier to make a project sustainable if you have a lot of money for the project. But often you don't – so you either have to fight for more money or to succeed with sustainability through good ideas and innovation. Either way, as in a military campaign, you must take it seriously and take your role seriously.

But even in difficult situations, there is hope and opportunity (Hanselman, 2016):

> Apply yourself to thinking through difficulties – hard times can be softened, tight squeezes widened, and heavy loads made lighter for those who can apply the right pressure. Seneca

There are always multiple paths to choose between to go through problems and projects – be creative. Often step one is to slow down and think the situation through. Can the times be softened, the entrance widened, and the load lightened?

A central point of stoicism is to think things through, not to give up, not to panic and flee, and not to make ill-considered attacks. Here, Seneca talks about the 'right pressure'. The right pressure to be applied in project management can be a series of conversations with a number of persons who, if they all do the right thing together and simultaneously, can solve the problems that the project has run into. Or it may be resolving a conflict, sharing costs, or solving a technical issue. But don't run head-long into it, rather reflect before you act.

Two quotes on character and willpower (Hanselman, 2016):

> For even peace itself will supply more reason for worry. Not even safe circumstances will bring you confidence once your mind has been

> shocked – once it gets in the habit of blind panic, it can't provide for its own safety. For it doesn't really avoid danger, it just runs away. Yet we are exposed to greater danger with our backs turned. After Seneca

Most projects contain a multitude of risks stemming from a wide variety of reasons, and therefore, projects can seem intimidating. It might feel like the project manager become a kind of 'punching bag'. People can become angry or uncomfortable when things go wrong, especially when there is a loss – no matter which organisation ends up on the paying side.

As a project manager you might get scared by the things that happen in the project or might happen. You might fear what you sometimes need to set in motion. There aren't many project managers with a few years of experience who don't know what it's like to wake up suddenly wide awake in the night fretting about how to handle problems arisen in a project. For many, it's 'just a part of the job'. Seneca advises confronting the danger, not hiding from it, and turning your back on it.

Some project managers develop the unfortunate pattern of minimising risk for themselves rather than for the project in order to have the least possible exposure to criticism if something goes wrong. This may be the case when a project manager raises a difficult issue but does not take steps to deal with the problems themselves. When things have gone wrong, they may indignantly say: 'I told you so!' Others tend to quickly explain that a supplier or project participant or anybody else have made a mistake. There are many ways in which our worries and anxieties lead us to try to protect ourselves instead of trying to save the project – and thereby ourselves.

It is less dangerous to take a more radical approach to your managerial responsibilities – grab hold and keep working until the danger is averted; at least in healthy organisations. In organisations focused on blame this might not be the case. Such organisations should in general be avoided as a workplace. However, even in healthy workplaces, it might still feel daunting to take on great responsibilities or take the responsibility for mistakes. Therefore, it is helpful to investigate a bit deeper into the word 'responsibility' and its meaning.

The term responsibility is comprised of 'response' and 'ability'. In other words, responsibility is about the ability to respond. Hence, the idea is that if you have the responsibility, you also possess the ability to respond. In Danish, the meaning of the corresponding word is similar. There it means that 'you owe response'. Both meanings are relevant. In any case, it does not mean 'blame-worthy' as many interpret it. Instead, it means you have the ability and the duty to respond. That is, if something goes wrong (and it usually will in projects of some complexity), you have the obligation and mandate to find a reasonable response – often in the form of a series of actions. This ought to take the anxiety out of the word and substitute it with leadership and courage.

The misinterpretation that lies latent when we use the word 'responsibility' as: 'Who is to blame when things go wrong' is not conducive to people taking responsibility. Imposing blame does not provide an answer and does not create a solution. A problem alone should not lead to blame. Instead, it is the lack of an answer that gives reason to blame. Therefore, it is much more helpful to see the management part as 'a duty and ability to respond' – to provide a solution. If you look at it this way, facing the danger of the risk is simpler and better than avoidance or pushing the blame forward to others. Because facing the danger is exactly what enables you to respond. You must get to a stage where you can react internally and externally and ask yourself (or colleagues) 'What do we do now?' With that understanding, it becomes evident that the greatest danger is to turn your back – because how can I hope to answer when I don't even understand the question? (Hanselman, 2016)

> You must stop blaming God, and not blame any person. You must completely control your desire and shift your avoidance to what lies within your reasoned choice. You must no longer feel anger, resentment, envy, or regret. Marcus Aurelius

Something magical happens when we stop judging each other and the world. But humans have an ingrained habit of judging everything, often with incredible speed and without sufficient reflection. When we practice suspending our judgement, we can observe more objectively what the truth is and how problems are related, and then we can act more effectively.

And to conclude this section on the stoics, let's just point to this quote, which seems to predict the sustainability agenda 1800 years ahead of time (Hanselman, 2016):

> What is not good for the hive is not good for the bee. Marcus Aurelius

It is a realisation of the moral theory of taking the community into account in all our considerations. Through this comparison, Aurelius brings the idea of this ethical practice to its actual meaning for everyone. It is an attempt to solve the problem of 'The Tragedy of the Commons', as described by Garret Hardin in 1968.

The problem in 'The Tragedy of the Commons' is that if everyone optimises their behaviour for their own gain, then the community's resources are used up or destroyed. For example, if fishermen compete to catch fish, the young small fish will eventually get caught, and suddenly you have a situation where the fish stock cannot be maintained. In the following years there are suddenly no fish or only a few small fish to catch. Thus, by optimising only regarding one's own needs, you destroy the basis for the whole – and thus also for yourself.

Marcus Aurelius reminds us that we live in a beehive. We are not alone. Thus, a fundamental dogma of sustainable project management is that everything is connected, and if you do things that are bad for the community, they are not suitable for you either.

THINKING FAST AND SLOW

'You cannot know this from conscious experience, of course, but you must accept the alien idea that your actions and your emotions can be primed by events of which you are not even aware'. This idea was written by Daniel Kahneman in his book 'Thinking fast and slow' (Kahneman, 2012). In 2002, Daniel Kahneman received the Swedish Riksbank Prize in Economic Sciences in Memory of Alfred Nobel, which is awarded together with the Nobel Prize and is popularly known as the 'Nobel Prize in Economics'. He received the prize for his lifelong research into how we think – especially how we don't think as clearly as we think we do.

The blind spot discussed by Kahneman is our ability to think in two distinct ways. He calls them systems 1 and 2. When we think about ourselves – who we really are – we basically identify with the process in ourselves that seems like a consciousness stream of inner speech or images running almost non-stop during our waking hours. Kahneman calls this process system 2. We easily overlook the silent system – system 1 – which runs speechless and unlabelled in the background with millions of pieces of information per second. Other words for this system 1 are our orientation system or our survival system. In comparison, system 2, our rational intelligence, runs with only a few pieces of information per second. Even so, system 2 runs with most of the attention.

The two systems (1 and 2) have a close seamless collaboration. Hence, it is not the case that we have two personalities or two independent agents operating in us; still it makes sense to spend some time understanding how the two ways of thinking functions differently. It can be helpful to be more aware of the two types of thinking and to be able to apply the two ways of thinking strategically. Most importantly, however, we need to think of this because if we are unaware, we can be misled into using system 1 for tasks where we should have activated the slower system 2. For example, sometimes it is an excellent idea to use our 'gut feeling' or 'intuition'. Without it, we would not be able to act in our everyday life; however, sometimes it is essential to pay close attention and activate our logical and analytical intelligence. The detours that system 1 can divert us to are tricky because we have no chance of detecting the detours unless we understand the mechanisms involved. There are numerous bias mechanisms we must be aware of to prevent ourselves being tricked.

First, let's feel inside ourselves how the two systems feel. First, try to solve the following tasks:

(1) $2 + 2 =$_____
(2) Don't sell the skin before the bear is _______
(3) Monday, Tuesday, Wednesday, ______
(4) What is $54 \times 17 =$_____

In questions 1–3, we use system 1 to answer, but in question 4, it is suddenly system 2 that has to work. You probably experienced a more significant mental strain in question 4. You had to exert yourself. This effort can be observed from the outside – for example, by the increase in the size of your pupil. Observing the pupil makes it possible in a laboratory setting to know when your system 2 is in action. It is also possible to measure by observing the pupil when you either reach a result or give up – because then the pupil will return to its original size. The use of system 2 can also be seen in increased blood pressure, a higher heart rate, and in your muscles contracting. It turns out that glucose metabolism is also higher while working in system 2. This is the reason that it feels like an effort.

System 2 tires quickly. When you're trying to work out complicated things, you can hardly do anything else simultaneously. If all information had to constantly go through system 2, your brain would quickly become overwhelmed and exhausted. But it doesn't have to. In total, 99.999% goes through the language-less system 1. System 1 is your best friend to help you get through the day safely. System 1 can even do fairly complicated things: a skilled driver can drive for hours on end without needing system 2 – unless you're tired. Then, system 2 must work hard again. When you are tired, system 2 must monitor the process so that you don't fall asleep. Hence, as soon as you can feel a mental effort, it is a good sign that system 2 is active.

System 1 is organised as a vast associative network – a multi-dimensional mind map. Your system 1 is fast and efficient at drawing on this network of associations. Its primary function is to assess whether 'the situation you are in right now' is normal. System 1 monitors whether what you are experiencing roughly corresponds to your understanding of the world. And the world does, after all, most of the time do as you expect. And therefore, you also know – most of the time – what to do next without thinking too much about it. Opening and closing doors, putting on shoes, and brushing your teeth don't require much reflection. As long as traffic behaves normally, driving takes up very little space in system 2, and instead, system 2 starts generating an inner stream of speech about anything and everything.

When, however, system 1 detects something abnormal, it calls on system 2. System 2 is then in charge of devising a solution to the strange situation. System 2 finds a way not to run down the cat on the road or calculates the value of 54 × 17. If an anomaly happens repeatedly – for example, a new and different word appears in the texts you are reading – then the new word is eventually added to system 1, whereby it has become a normal thing. From there you have easy access to the new word. Similarly, you need to expose your system 2 to a lot of knowledge about project management and sustainability to automate it, so that it doesn't require a lot of effort all the time. However, learning always goes through system 2's effort – that's why it always feels so 'hard'.

Kahneman spends most of his popular book describing how system 1 sometimes leads our thinking astray. This can be, for example, by an optical illusion or bias (systematic error) in our thinking; for example, when we err by associating foreigners with crime, women with weakness, or men with top management.

Our associative network has emerged from what we have experienced and have been told and observed (repeatedly), both consciously and unconsciously. System 1 to a great extent shows the owner his/her world as he perceives it – if he/she did not question the input, then the associative network will just represent the world as it looks with whatever flaws the person have not yet weeded out. This is why there are some male chauvinists and racists who will claim that they are not. They are simply blind to some of their beliefs – the beliefs are built into the lens through which they see the world. And that's also the reason why it's sometimes so challenging to understand people who think markedly differently. The inner network is a kind of hyper-advanced model of the world and how it is presented to us. The system has a way of overlooking errors unless you pay close deliberate attention. Errors that are not critical for our survival tend to persist. If only we could take the model out and look at it and find the errors. The problem is that it cannot be taken out. It is hardwired in us. The world therefore looks as it presents itself to us (and not as it really is). An optical illusion is almost impossible not to see wrong, it takes an enormous effort to see it for what it really is – even when we know it.

An example of a problem that this can create is that we interpret an unusual sequence as more unique than it really is. System 1 is a poor statistician. It sees patterns where there are coincidences. There would be an outcry if a lottery series were drawn that was 1, 2, 3, 4, 5, 6. But this sequence is just as likely as 1, 7, 28, 31, 56, 77. It is surprisingly difficult by imagination to generate anything that is random. The associative machine is simply too powerful. This is why relatively simple mathematical methods can detect fraud in accounts where attempts have been made to replace real numbers with made-up numbers.

Research has found a wide range of ways our brains cheat us. Here is a list of eight types of bias that may be of particular importance as project management working with sustainability:

(1) **Availability bias**
We tend to believe more in the examples that first come to mind – even if they are less likely.
For example, if we hear a lot in the media about immigrant criminality, we will tend to make the connection that immigrants are more criminal than they are in reality. Regarding project management, we may tend to quickly draw on lessons learned from historical projects where a situation/problem was somewhat similar and then adopt the same approach. It requires something of us to truly appreciate the situation as new and unique. To see it for what it is and find the best solution available. Another example is what is called 'carbon tunnel vision' within the field of sustainability. Because we are discussing CO_2 so much, we may not realise that there are other climate gases, that there are other climate solutions than reducing CO_2, for example, planting more trees and that sustainability is a much wider concept than only climate change (refer back to the chapter on planetary boundaries and doughnut economy).

(2) **Groupthink**
Because of our natural aversion to conflict, we tend to value conformity too much. This often leads us to too easily adapt our views to maintain group harmony, which can be seen in various contexts. However, difficulties can arise when the group lacks the ability to think outside the box or when a few members dominate the agenda, leaving others hesitant to challenge it – even when it's appropriate and ought to be done. The influence of the person speaking up first can sometimes sway the entire group in a particular direction. This phenomenon is also connected to false consensus bias, wherein a project manager mistakenly believes there is greater agreement on chosen strategies or decisions than there actually is.

(3) **Favouring the in-group**
In many projects, there is a 'we are the good guys' feeling. This can cause problems between members of different organisations working on the same team. A common issue is that two subgroups from each their different organisations have their own understanding of whether a cost was unforeseen or due to lack of care. Or whether a cost is within or outside the agreed scope of the contract. In the worst-case scenario, such discussions can develop into resource-intensive disputes or costly arbitration. In a team, it is crucial to establish a sense of 'the in-group includes everyone on the team', even – or especially – across organisations. Maintaining cohesion is an important task throughout the project to avoid unnecessary and resource-intensive conflicts.

(4) **The Dunning–Kruger effect**
The Dunning–Kruger effect is that people who know little about a subject tend to have a tendency and an inner sense that they know more than they actually do. This is what makes many people suddenly feel that they can make very confident statements about, for example, what to do during a Corona epidemic. Surprisingly, there is also the opposite effect: people who know a lot are more cautious and less confident. The common explanation seems to be that they understand all the ways they can be wrong. A famous Charles Bukowski quote goes: 'The trouble with the world is that the intelligent people are so full of doubt, while the stupid are full of confidence'. The Dunning–Kruger effect is often seen in projects where somebody in the project environment is suddenly full of good advice or being very judgemental, even though – or because – they do not know all the contexts and dilemmas of the project.

(5) **Confirmation bias**
We have a strong tendency to favour information that confirms what we already know, think, and believe. Although information that goes against our worldview tends to go under our attention radar, we simply don't hear it. This is probably because it takes 'system 2' a lot of energy to change our worldview. This can have serious consequences in projects if we don't realise something is off, a small mistake may in the worst case topple the project. Often when something has gone wrong, we feel that someone has already mentioned the problem to us, but not 'loud enough' for us to take it seriously and react. Confirmation bias is a massive problem in sustainability skills because we prefer to solve problems the way we always have, so confirmation bias can discourage innovation without us realising it.

(6) **The law of triviality**
This bias gives us a tendency to pay disproportionate attention to straightforward well-known and well-understood issues while avoiding dealing with issues that are complex, confusing, and not as well-known to us. This may lead to huge waste of time in projects; especially, where one or more people tend to have self-promoting tendencies. This can result in more time and resources being spent on, for example, deciding how the bicycle shed (simple) should be designed than on the choice

of building materials based on their environmental impact (complex). Additionally, it may lead to serious and complex issues not being dealt with in an appropriate way.

(7) **Sunken-cost bias**
When we have spent a lot of resources on a project, we tend to continue even if the business case for the project changes in the negative direction – or we become wiser in other ways. We find it hard to bear that the money and effort already put in is wasted. The advice 'don't throw good money after bad' tries to help us avoid this bias. Closing projects where a lot of resources have been spent takes a lot of guts. Still, remember to celebrate when you do it, because you may just have saved a lot more.

(8) **Authority bias**
We tend to trust and be more influenced by the opinions of authority figures. This happens, for example, if the chair of the steering group thinks in a certain way. It is easy to think that the authority figure knows more, but in projects this is often not the case. It is important for the project manager to challenge opinions in the steering group they are not fitting to the context of the project – which the project manager mostly knows better than the steering group.

Failure to activate system 2 thus gives rise to number of everyday problems. First, it is necessary to recognise where bias occurs – otherwise, we cannot 'catch ourselves in it'. In the same way as when a tired driver has to use her system 2 to monitor that she does not fall asleep, it is an effort. It is called 'to pay attention', for a good reason – you actually have to pay attention and energy to system 2. It's an expensive system, and you only have a limited amount of attention available.

An additional important point to know is that when you are relaxed and happy, you are more likely to use system 1. However, when you are in a slightly more difficult and strained mood, you tend to be more attentive and suspicious and thus better able to avoid mistakes, simply because you then tend to invest more system 2 attention to what you are doing. So, from that point of view, it may (unfortunately) be better to be tense than happy when doing projects. However, don't disqualify system 1, your intuition, experience, and gut feeling, but rather set multiple alarms for when you should call on system 2 and create a practice of regularly engaging system 2 through active reflection – both alone and in the group. You may want to ask the group: 'Are we doing this or that in the best way or are we overlooking something?' It is not easy to unlearn old habits, principles, and behaviours and to learn new ways of seeing the world.

'The ability to observe without evaluating is the highest form of intelligence' is words of wisdom attributed to Krishnamurti. You might think it was the other way around. That it was evaluating that was the highest form of intelligence – we evaluate all the time! That's pretty much what being a project manager is all about! But it's an interesting function to try to train to hold back evaluation. Observing a thing and withholding the evaluation is not as easy as it sounds – because system 1 is fast and automatic. Withholding judgement is often an excellent way to gain new insights. That's what reflection is all about – observing for a time without making up your mind.

So, if we return to Kahneman's quote at the beginning of this chapter: 'You cannot know this from your own conscious experience. Instead, you must accept the alien idea that your actions and emotions may be initiated by events you have not even realized or noticed'. Interestingly, all these bias problems occur below the radar of our awareness. It has taken concerted effort and research to realise the described bias mechanisms. The only way to guard against the system 1 mistakes is to reflect on our projects. Setting aside time to review the projects with our 'system 2' at regular intervals and considering what types of bias we might have fallen for – use the above list as a checklist.

Many project managers think that monthly project status reporting is a chore. But if you use the opportunity instead to reflect more deeply upon your project, the time may be very well spent. To succeed requires knowing where bias typically occurs, and thus, to have an overview of the situations where the brain plays tricks on us. Otherwise, you may be like a child falling for even the simplest illusions of the magician. Generally, bias is easier to spot in others but almost impossible to spot in yourself (or else it wouldn't be a bias). So, pay attention on yourself, your thinking, and your underlying assumptions. Just because your easily accessible beliefs get you through the day more or less safely doesn't mean you can run a big project on gut feelings, think carefully.

WHAT IS THE PRIMARY PRINCIPLE OF YOUR LEADERSHIP?

Part of the managerial challenge is to forge a way that the heart can support the brain and the brain can support the heart. We speak of management and leadership as two branches of the same thing: management with the head and leadership from the heart. Both are important for success – your own and your team's. It is important to train ways where the two support each other.

Most theory on project management has focused on rational approaches (the management part), primarily discussing methods and tools. It seems that the 'heart-centred leadership' aspect of project management has been taken less seriously for project managers than in line management, where it has gained prominence in recent years or decades.

Taking leadership seriously means prioritising personal development. Personal development is healthy for everybody: the project manager, those being managed, and the surrounding organisation(s). The project manager matures as a person. Those under management will receive increased attention, empathy, and a more comprehensive understanding of their needs. The organisation reaps rewards from well-functioning project groups, as they exhibit heightened precision and an increased ability to maximise the potential of each project. As a result, this generates greater value for the organisation.

There are many different paths to 'leadership from the heart' and personal development. The philosopher Søren Kierkegaard provides us, for example, with an interesting reflection based on his wanderings in the city of Gilleleje in his home country Denmark:

> The crucial thing is to find a truth which is truth for me, to find the idea for which I am willing to live and die. What good is it if I find a so-called objective truth, even if I worked my way through all the philosophers'

> systems and could, when necessary, give an overview of them all and show where each set of ideas was inconsistent – or if I could develop a unified theory of the state by combining a total coherence out of details from many places.
>
> If I could thereby construct a world in which I did not live, but which I merely held up for others to see – what was the use of me being able to develop the meaning of Christianity. And the meaning of judgement and to be able to explain many individual phenomena if for myself and my life it had no deeper meaning?
>
> This is what I needed to lead a completely human life and not merely one of knowledge, so that I could base the development of my thought not on – yes not on something called objective – something which in any case is not my own, but upon something which is bound up with the deepest roots of my existence, through which I am, so to speak, grafted into the divine, to which I cling fast even though the whole world may collapse. This is what I need, and this is what I strive for.

Take the Christianity out of the quote (or keep it, as you please) and it describes the ultimate in 'leading from the heart'. '…bound up with the deepest roots of my existence', means look into what is the most important thing to you? Ask into the deeply important. And look at the subjective. This is a question that is not easy to answer. You may never reach the point where you have perfect clarity in your answer. But looking inwards and attempting to answer that question is central to strong leadership.

The purpose of creating a discipline called 'sustainable project management' is not to have a line-up of similar project managers with similar competencies – quite the contrary. The goal is to develop project managers who are so confident and self-aware that they have the courage to stand in full height, on their own two feet and fully embrace their authentic selves. We hope for you that you will be well-reflected about yourself, that you will use your strengths actively, and that you unashamedly will ask others for help where you have weaknesses.

Pia Søltoft is a learned Danish Kierkegaard scholar, in her book which title translates to '10 things leaders can learn from Kierkegaard', she outlines 10 principles from which to develop leadership from the heart (what Kierkegaard calls in the quote: 'the deepest root of my existence') (Søltoft, 2018).

The fundamental concept we can draw inspiration from is that as leaders, we should select a personal ideal – 'a truth that is true for me' – or, in other words, 'to will only one thing'. The challenge that Kierkegaard poses us with, is to find the Idea (capital I), the Truth (capital T), the Ideal (capital I) against which we will measure our leadership behaviour. Kierkegaard would extend that and say the ideal against which we will measure our whole life. In this way, in the end, there can be no greater difference in the ideal between 'the man of work' and 'the man of leisure'. If a consistent and genuine demeanour is not maintained between your professional and personal life, it may be seen as a lack of authenticity.

'To will only one thing', that is, to have one ideal leads to the difficult question: if I had to choose just one ideal to be the yardstick for the rest (!) of my life, which ideal would I choose? It must be an ideal that promotes the good for me, my relationships, and the whole. Examples could be: I will do my utmost in all relationships to be fair, prepared, attentive, non-judgemental, caring, or conscientious. In the context of this book, it is relevant to play with and perhaps later work with the idea of always working for sustainability.

To elevate this to the overarching ideal raises a whole series of reflective questions:

- What do I mean by sustainability?
- Which dimension should have priority?
- How do I translate it into something practical that delivers real results in the world?
- Should I be a sustainability apostle who never talks about anything else?
- How do I work on my communication so that I am heard?
- How do I make my life function where it is difficult to make sustainable choices?
- What do I do if it is not possible to make my projects live up to my ideal?
- When will I chose to adapt to the non-sustainable circumstances and when will a compromise weigh so heavily that the project start eating away at my conscience?
- How can I improve on the next project compared to the last one?
- How do I recognise the opportunities? How do I ensure that I am sustainable and not end up in burnout?

Kierkegaard provides a series of instructions on why and how to live one's life with an ideal of one's own choosing.

The first step is to distinguish between freedom and necessity. Kierkegaard does not subscribe to the idea that 'anything is possible'. Just as we experience it every day in many ways in project management, when we, as project managers, embark on a project, there are always necessities around a given project. A project is significantly influenced by factors such as the available team members, the composition of the steering group, and the actions of prominent stakeholders' representatives. Most projects arrive at our doorstep with an entire cast of characters filled before we even begin choosing and forming the project. And the people can be exchanged – outside of our control. As a project manager, you must get the best out of the cast of characters. Projects also usually come with a financial imperative. Often the financial imperative can be negotiated – at least to a point – a negotiation that takes place between several parties. Careful management of finances, including goals, deliverables, success criteria, timeframes, risks, and stakeholder involvement is crucial for success. However, even with exceptional expertise, most projects will encounter financial necessities.

In other words, projects are always bound by a lot of 'necessities'. But between these necessities, there is always room for free choices. To many people, this room is almost invisible, you will have to actively look for it. As a project manager, you are constantly managing this room of freedom. Through time management, you can choose to spend time on the schedule, follow-up, stakeholder management, risk analysis, or talking to a project participant who is struggling. You can choose to work on the long-term trajectory of the project or the short term. You can be disciplined about project management or not. What is important is that you recognise the gap between the necessities of the project and your freedom to choose how you want to manage it. And that you grab that freedom and use it well.

But how do we judge if our room for free choice has been well administered? This is where your Ideal or central Idea is important. If your ideal is to do your utmost to be a sustainable project manager, then it is that decision that must guide you to prioritise your actions in your space of free choices.

Deciding on your ideal is not a straightforward process. You can probably feel the difficulty and heaviness of determining the ideal that will guide you for the rest of your life. It's a big decision. Therefore, we tend to want to put it off until later … and later … and later. Kierkegaard would claim that it is precisely our sense of 'there is still time' that is one of the greatest dangers in a human life. Time carries us without pause on our journey through life towards our death. But when will we make our crucial decision? Will we choose while we can still live a life according to our ideal, or postpone it until our final hour when our freedom is exhausted? What do I want to do with my life and what do I want to do in my life? How do I want to use my freedom? What meaning do I want to give to my life? In this way, our ideal is not a narrow straitjacket but a gift to us. A chosen meaning or solution to the task that life has given us. To get in touch with this important question in ourselves helps us to make the decision.

Once we have made up our mind, the next problem arises 'keeping the decision fresh'. There are plenty of reasons that today is not the day to act, we can postpone until tomorrow or the day after or … and suddenly we realise that we almost forgot our idea. Kierkegaard rather harshly calls this perpetual procrastination a 'cowardice' in us. A cowardice that we all have and thus need to stride against – a bit like cycling against a headwind. We must overcome the headwinds again and again, day after day. When things go wrong in a project, upon closer scrutiny, we realise that we have been at least part of the reason for the problem – often in the form of 'a sin of omission'. In this way, there is often much faster feedback for errors in project management than in life. Every decision matters. A single misstep could derail the entire project. This underscores the significance of being clear on our ideal and keeping it fresh every day – ready to start all over.

Another important reflection is: am I managing my ideal fanatically? Kierkegaard does not ask for fanaticism from any human being. An ideal should, therefore, not be held in a clenched fist but rather be so soft that it can wind its way through our lives. It is inevitable that sometimes you make mistakes. You choose the wrong goals or take decisions and actions that don't work out. It is important that you do not keep yourself in a situation where you continue down the wrong path to avoid losing face. That would be fanatical. And sooner or later, it will go wrong (trust this)! It is, therefore, essential to be able to change your mind, to be able navigate adeptly to feedback on actions and directions that are not working well. The ability to detect and understand feedback and react wisely is central to leadership. If you let pride or fanaticism, get in the way, the project will soon be in dire trouble. Additionally, consider that it is sometimes better to move in the wrong direction than not to move the project at all – provided you are prepared to navigate, that is, take feedback and change direction. So, don't be hard on yourself (or others). It is essential for project managers, especially in a project department, to adopt a forgiving and growth-oriented attitude towards mistakes and opportunities for improvement. In that way you ensure that everybody can take responsibility for mistakes without shame or ridicule.

In the best-case scenario, you tell the project group and the steering committee openly about mistakes and explain how you plan to change the direction going forward – that is taking true responsibility. By standing up, assuming responsibility, and altering your path, you generally enhance the trust others place in you. They can see your authenticity and your commitment to managing the project in the best possible way. If you on the contrary have a tendency to keep problems hidden, prolong the inevitable and don't own up to your mistake, people will eventually lose trust in you. You can create a noticeable shift in your own mindset, for example, by adopting a guiding principle such as: 'I am the kind of person who takes responsibility for mistakes and corrects them'. People will sense this and trust you more.

A meaningful way to use your ideal is to remember that the end and the means are equally important. That your ideal applies to both the big issues and the small issues – even to the almost insignificant. The situation can be likened to an archer who, upon hitting the target, is celebrated for his achievement. However, if the hit was a lucky shot, the praise quickly fades. Thus, the means – in this case, honing his aim through practice, is equally crucial. Similarly, investing time in reading a book like this holds great significance for your management skills – even if nobody will know you invested this time. Reading books on important and relevant topics for your overall ideal help you clear your mind and reflect on important questions like: how do I 'aim like an archer' in project management, and how do I keep improving?

We are not pre-programmed to follow a predetermined route. In the midst of all the necessities of life, resides freedom. The freedom of choice. But we need to use this freedom in a conscious and wise manner and as such freedom is also a kind of obligation. We can misuse it – and this claims Kierkegaard is a part of our existential anxiety. Therefore, anxiety (to some degree) could be said to be a constant companion on the path of self-development. Anxiety, unlike fear, does not have an external danger as its cause. Instead, Kierkegaard portrays anxiety as a manifestation of freedom, where one's choices hold significant importance in shaping the person we are to become. This involves living according to your ideal and making conscious efforts to become your true self. In such a perspective, you cannot dismiss your current job by, for example, shrugging your shoulders and say 'oh, whatever, it's just a job'. Our work as project managers matters greatly – also for our own becoming.

MAIN POINTS

Everyone knows the importance of thinking things through and of thinking clearly. But research has shown that it is surprisingly unclear what to think about and how. If sustainable project management is important to you, then this chapter provides you with four approaches to strengthening your thinking and thus accelerates your path towards becoming a sustainable project manager able to deliver sustainable projects – and eventually be the co-creator of a sustainable future. The four approaches in this chapter all starts with you: what is really important to you, and how do you translate it into 'in the real world' action?

Working on one's ethics is a systematic way of thinking about what it means to 'be a good human being'. And when you delve into ethical questions, they are filled with surprising and hidden dilemmas. Dilemmas that tend to be much more prevalent and comprehensive than our day-to-day life seem to suggest. There are ethical dilemmas with clear yes or no choices, there are dilemmas with scales where choosing an appropriate level is key. There are dilemmas where two or more oppositional considerations need to be balanced. There is the question of ethical wingspan, that is, how much of the whole do you take into consideration in your ethical decisions? How transparent can you be in your choices, and does it matter how your actions are perceived and judged by people around you – people who may not know the full context? What about the knowledge you don't seek – where are you guilty of 'wilful blindness'? What actions today would make most impact if you were to become more ethically informed?

Stoicism helps to frame our choices and provides a framework for how we can think about them. Three main themes are discernment, discipline of action, and strength of character. These three main themes are of great importance for your engagement in the role of sustainable project manager. How you are perceived and how you perceive yourself matters for your impact. Applying a stoic outlook on project management will help you prioritise and give you some clear directions on how to lead your projects and life in consistent integrity.

Kahneman's research has become very popular these years. You'd have to look hard to find a Nobel laureate whose theories have been so widely read. His point that we are deceived by ourselves without realising it is both disturbing and important. Once we understand the problem of how our brains trick us, we can recognise the problems and avoid stumbling into them. We must know where bias in the design of our thinking will trick us. It is like optical illusions, until we realise the tricks, we remain victims of them. And it is hard work(!) even when we know something is an optical illusion, it is difficult to see the truth behind. Kahneman's theories of bias thinking are important to all of us.

With Kierkegaard's existential reflections, we come right up to the biggest of pictures, where we decide what ideal (capital I), we want to live by: our vocation, our purpose in life. When do we decide this, and how do we commit ourselves to the ideal? Kierkegaard makes a great companion for any leader as a source for deep reflection.

YOUR REFLECTIONS

To get acquainted with the ideas of ethics try to review yesterday for ethical dilemmas with the following diverse approaches:

Can you think of an ethical dilemma you encountered yesterday?
What options – if any – were you aware of when the dilemma arose? What options can you see now? What did you choose? Why did you choose it? Would you choose it again today?

How much of the whole do you take into account in your moral and ethical considerations?
When do you prioritise yourself first? When – and how often – do you choose your family, your friend, your organisation, your company, your country, the whole world? In other words, how broad is your ethical wingspan?

What information do you tend to avoid because it would put you in an ethical dilemma?
How do you react when fundraisers from Red Cross, World Wildlife Fund, Greenpeace, and so on want to speak to you? How do you react when you meet beggars at home or abroad? What do you think about people who are struggling in one way or another – poverty, obesity, handicap, race, and so on? How often do you choose to take time to understand the situation and context of somebody else more deeply?

It is hard to admit – even to recognise – where we have been wrong. But we all make mistakes on a daily basis. So, try to see if you can spot episodes in your life where you have been wrong with these biases: availability bias, groupthink, clique favouritism, the thinker's effect, confirmation bias, the law of triviality, sunk cost bias, and authority bias. If you can't think of at least a few episodes per type of bias, maybe you need more time for reflection?

Stoicism has many good quotes. Here are a few more that you can try to 'unpack', so that you can gain an understanding and advice for yourself:

> Don't return to philosophy as a taskmaster, but as patients seek out relief in a treatment of sore eyes, or a dressing for a burn, or from an ointment. Regarding it this way, you'll obey reason without putting it on display and rest easy in its care. Marcus Aurelius

> Above all, it is necessary for a person to have a true self-estimate, for we commonly think we can do more than we really can. Seneca

> Philosophy does not claim to get a person any external possession. To do so would be beyond its field. As wood is to the carpenter, bronze to the sculptor, so our own lives are the proper material in the art of living. Epictetus

In ancient Greece, there was an idea of the 'philosopher king', particularly in the works of Plato. Plato believed that a good leader is also a philosopher. That is, someone who have studied and gained wisdom through philosophy in order to rule justly and wisely. Philosophy examines fundamental questions relevant for how to lead our lives and is one of the best leadership development tools available.

Following Kierkegaard's model, consider finding 'the truth that is true for you' – something that is connected to 'the deepest root of your existence' – 'even if the whole world collapses'. Why are you postponing this decision? Is it not relevant for you? And then when the decision has been taken: how do you practice it?

Have you taken your daily dose of philosophy?
Spend some time with the stoics or Kierkegaard – or venture into other philosophies. They are great for helping you question the things you thought you knew for sure already – and we all need that.

READ MORE

Ryan Holiday has written several books and recorded YouTube videos on stoicism. 'The Daily Stoic' is an excellent book to continuously work with stoic quotes (Hanselman, 2016). His book 'The Obstacle is the Way' describes a good approach to looking at problems differently (Holiday, 2014). In addition, there is joy in going straight to the source and reading, for example, 'Meditations' by Marcus Aurelius (2023), or 'Stoic Rules of Life by Epictet' (Epictetus, 2023). The chapter on stoicism is greatly inspired by Ryan Holiday's work.

Ken Wilber's book 'Integral Life Practices' has a chapter on ethics, which has guided the introduction of ethics to a project management perspective above. It is generally a book to learn from in many different areas (Ken Wilber, 2008).

Daniel Kahneman's book 'Thinking – Fast and Slow' is a popularised version of the studies that Kahneman conducted together with researcher Amos Tversky, reviewing the many different laboratory experiments that have revealed how unreliable our brains sometimes are (Kahneman, 2012).

Pia Søltoft's book '10 Things Leaders Can Learn from Kierkegaard' is an exciting dive into essentially straightforward questions about what kind of project manager you want to be – and what kind of person you want to be (Søltoft, 2018). Regrettably, the book is only in Danish; however, numerous international books on Kierkegaard are available, such as for example, 'Kierkegaard: An Introduction' by C. Stephen Evans (2009) or Life Lessons from Kierkegaard by Robert Ferguson (2013).

In addition, there are plenty of other books to get started on the four topics. It is also meaningful to work with other philosophical approaches. A simple and easy-to-read book that gives a good overview of the development of philosophy is 'Sophie's World' by Jostein Gaarder (2003). It was originally meant as an introduction for young people, but it is a good starting point adult as well. You may use the book for choosing your favourite leadership philosophy.

doi: 10.2166/9781789063349_0143

Chapter 5

Be aware

Do it as well as you can until you know better. When you know better, do better

Maya Angelou

In the project manager's universe, situations arise all the time that could have had a better outcome if the parties involved had considered whether they could have adjusted their perception of the situation and their behaviour. In the day-to-day running of projects, conflicts, delays, and cost overruns often arise that can be attributed to inappropriate behaviour or a lack of consideration for the project.

We think and do the same things according to a pattern that we repeat over and over in many situations. We are creatures of habits and have a strong tendency to let the autopilot run the show. However, the brain's mental processes receive input from the senses and regulate physiological and biological processes within the body down to the cellular level. This rapid feedback loop ultimately shapes our physiological and emotional states, as well as our behaviour. If you wish to change your behaviour, begin by examining your mental patterns and being mindful of your mental and emotional processes. Initiate change by experimenting with new thoughts, choices, actions, experiences, and emotions. This will affect your personal reality and biological aspects such as nerves, internal chemistry, genetics, and potentially even health. In many cases, altering habitual actions is crucial because if we always do the same thing, we will eventually get the same result. If something fails, approach it differently next time you encounter a similar situation. Adopting this mindset can be challenging amidst a hectic daily life as habits govern a significant portion of our bandwidth. However, this mindset can bring about meaningful changes in your everyday life as a project manager.

In this chapter on awareness, we focus on the sensory system and the role of beliefs in guiding our behaviour. We believe that future sustainable project managers need insight into key aspects of human behaviour, an understanding of its underlying factors, and the ability to adjust behaviour to achieve desired outcomes. Recent knowledge refers to the heart as 'the emotional brain'. In this chapter, we urge you to connect with your heart as an essential aspect of experiencing emotions.

One challenge is that many of the models and methods we describe are basically a simplification of reality. Mental, physiological, and emotional processes are complex, but still having a simple model may be helpful. A lot of research is going on about awareness these years – especially in neuroscience. We refer to elements of this research where we deem that basic descriptions may be helpful.

USE YOUR SENSES

We process information through our senses. Continuously bombarded with sensory input, we construct a personalised map of beliefs, interests, and thoughts, leading to our unique perception of the world. The subconscious receives over 10 million stimuli from the environment every second. To focus on our sensory system doesn't necessarily mean sensing more; rather, it involves becoming more conscious while sorting through these impressions. By improving our recognition of differences, we can make better choices and heighten our sensory awareness. The sensory input is the input to the brain, which uses the electrical impulses and hormones in the neural pathways to analyse, judge, and make decisions.

An example of a critical hormone is oxytocin. Oxytocin is also known as the 'love–hate hormone'. It is released in the body by hugs and prolonged eye contact and forms part of a reward system in the brain. Oxytocin gives a feeling of happiness when released and can be important in long-term collaborative relationships where individuals form close bonds. For example, in projects, you might experience such moments when everything is successful or when a conflict has been successfully managed. Oxytocin is characterised by the fact that once people have experienced it, most people want more of the same feeling.

The sensory system receives information and impressions from the environment through the seven senses: sight, hearing, taste, smell, touch, balance, and body or movement. Even when you are asleep, your sensory system is at work. We primarily store our memories visually, aurally, or kinaesthetically. People have different learning styles with different preferred senses. Awareness of one's preference provides a key to self-understanding and hence preferred learning style, working method, reactions, and behaviour. Knowing a person's preferences may also improve effectiveness in communicating with them.

The primary sensory systems are not mutually exclusive, but it is difficult to focus on multiple sensory inputs simultaneously or to identify the order in which they enter our consciousness. Some individuals prefer to see and hear, whereas others need to feel and hear. To identify your personal pattern, consider your last vacation or a specific word and try to discern whether your initial impression was an image, sound, or feeling. Preferences provide a simplified version of reality, and although an individual may be predominantly visual, auditory, or kinaesthetic, the truth is that our experiences are a blend of all our senses – with one or two taking prominence.

Eye movement is related to the way a person processes information and which parts of the brain are activated (Seymour, 1997). A visual person will tend to look upwards. The eyes typically move to the left and right side if the person is recalling sounds. However, a kinaesthetic person will often correct eyes downwards to the right and to the left. Therefore, one way to recognise the primary sensory apparatus of others is to record their eye movements.

When communicating with people, there are several things you can do to make it easier for them to engage and comprehend according to their preferred style. When speaking to a team consisting of several people, it's wise to adjust your communication style to accommodate all three types of perceptual styles.

People who are primarily visually orientated

You can recognise people who are primarily visually oriented by their body language, where eye movements and gestures will be upwards. Visual people tend to look up to either the right or left when receiving information. 'Visual people' often speak quickly and in short sentences. They describe things in broad strokes, for example, with the headline of the internal image they have in their mind's eye, leaving the listener to fill in the details. Visual people typically use words such as 'imagine', 'see', and 'perspective'.

The visual person typically has attention to colours, light/darkness, how things and people look, and a preoccupation with aesthetics. A distinction can be made between external vision, what the eye sees in reality, and inner vision, which is the imagination of the inner gaze or visualisation. These individuals typically experience conscious and unconscious visualisations on an ongoing basis, and the inner gaze is used as navigation to help understand and act in everyday life.

Visual learners learn best by visualising things and getting an overview first. If the preferred sensory system is visual, it is a good idea to communicate to the person in language containing visual references. Tips to help visually orientated people include:

- Feel free to use visual aids to communicate, for example, charts, diagrams, drawings, graphs, and maps.
- Use concrete examples or visual demonstrations.
- You can encourage visualisation with sentences that start with, for example, 'imagine that you …'.

People who are primarily auditory orientated

Auditory learners primarily utilise their sense of hearing and are highly perceptive to sounds. They can be easily distracted by external noises. These individuals often have a strong affinity for music, with their body language mirroring that of an attentive listener, evident by a slightly tilted head. Their eye movements tend to be focused straight ahead or to the sides.

Auditory people may often use expressions such as 'I hear what you say' and 'That sounds good'. In the project, you will experience auditory people as those who want to 'have their say' and need to discuss things thoroughly. People who use the auditory system a lot also learn most efficiently talking things through. They can easily reproduce long sentences verbatim. They often speak in long sentences, telling things as they come to mind, that is, as they hear them in their inner ear. The inner monologue is typically very dominant in their understanding of everyday life. It is as if they are almost telling themselves what is happening. Good advice to help auditory-orientated people can be:

- Use conversation and discussion, auditory-orientated people need to talk things through and hear different perspectives.
- Use assistive technology with sound, such as podcasts, music, lectures, or loud reading.
- Acronyms, doctrines, mnemonics, and the like are often helpful.

People who are primarily kinaesthetically orientated

Kinaesthetic means thinking in terms of bodily sensations, such as how the clothes feel against the body, how the chair feels to sit on, how something feels in the body, and so on. People are often good at sensing moods and how they and others feel. Body language is more profound than in the other representational systems; eye movements are downwards, and they tend to speak slowly and with emphasis.

Individuals who are kinaesthetically oriented typically appreciate physical recognition, like touch. They learn best through experiencing and feeling; it helps when they can empathise with a main character in a story or receive detailed explanations. Kinaesthetic language focuses on the perception of feelings and moods. Kinaesthetic language therefore focuses on description of feelings and moods.

In projects, kinaesthetically oriented people often exhibit thoughtfulness because they take the time to process their experiences by integrating available details and connecting with their emotions.

Tips to help kinaesthetically orientated people include:

- Try to create situations where it is possible to experiment directly – preferably hands on.
- Make sure there is room for movement during the meetings, change position between stand, sit, walk, and so on.
- Role-playing can also be a good method to use to increase the possibility of direct empathy with the situation.

Becoming aware of one's own sensory preferences and developing the ability to recognise others' preferences can be achieved by observing behaviour. We have found that differences in individuals' sensory systems can lead to misunderstandings or misinterpretations of others' actions and patterns. This is seldom beneficial, which is why it is crucial to be mindful of both your own and other people's senses.

You create inclusiveness and understanding by stimulating all senses in meetings and other settings – regardless of your own preferences. For example, when discussing solutions to a problem with your team, you may be aware that visual people prefer to see drawings or process diagrams showing the problem; auditory people want you to talk about the issue and allow them to share their own version, whereas kinaesthetic people may need time to feel and search for experiences from similar problems. The participants in your team and in your meetings will feel better recognised and understood if you add this attention to different preferences.

A great way to hone your skills in identifying sensory preferences is by watching interviews on video. By observing, you'll be able to discern the sensory inclinations of the interviewee and improve your ability to recognise them in others. With this knowledge, consider your coworkers individually. Can you determine their preferred senses? How can you engage their sensory preferences to enhance understanding between you? What adjustments can you make in your future conversations with them to cater to their preferences more effectively?

The best thing you can do to improve your ability to create a good dialogue is to start recording your own form, messages, timing, and so on. And at the same time, record the response you get back from the recipient. Then you can adjust, if necessary, the form, content, timing, and so on, of the dialogue. Learning takes time, and so does a change in your communication habits. Eventually, you'll notice an improvement in the responses you receive.

ACTIVE LISTENING

Among our senses, hearing is the first to develop and the last to fade away as we near death. Does this imply that we rely on hearing for a significant portion of our development? Many misunderstandings and conflicts in projects stem from a lack of listening to one another. As others speak to us, our brains seem to take control, generating sentences, words, and images to convey our thoughts that we need to express. Few people manage to listen all the way through somebody else's talk. We have two ears and one mouth for a reason. A lot would be different if our communication were geared to that perception. If you persist talking (too much), it prevents you from listening to others and to your own inner voice. Even our inner talking needs occasional to fall quiet.

Back in 1957, psychotherapist Carl Rogers and psychologist Richard Farson introduced the concept of active listening (Rogers, 2015). The core principle of active listening involves being fully present in the moment and completely focused. This requires setting aside personal attitudes and values during a conversation to truly engage and understand the speaker.

Active listening demands awareness and patience. Particularly when the speaker holds views contrasting your own, it is essential to be respectful and patient in comprehending their communication style. Letting go of inner judgement and personal biases can significantly enhance your active listening abilities.

Often, we engage in automatic and unreflective listening, where we are preoccupied with other thoughts while listening. This type of listening is superficial and focused more on our inner dialogue than the person speaking. We might also listen without concentrating. Often, we tend to plan our next response. Sometimes, we form our opinion in advance, leaving us seem uninterested in the other persons words and merely waiting for our turn to respond without engaging in genuine conversation. One technique to refocus our attention on listening is to ask the speaker to repeat or clarify their statement, which can enhance the dialogue and improve our understanding.

Greater participation in a conversation is achieved by being more aware of listening, being dedicated to understanding what the other person says, and asking clarifying questions. A deeper comprehension occurs when we set everything aside, become entirely present, and listen intently to the speaker. It allows us to pick up on nuances in speech, emotions, word choices, and imagery. This kind of communication happens when fully engaged in the present moment, without any reservations or preconceptions, and maintaining complete focus. At this level, external distractions and internal dialogues fade away, allowing for our undivided attention on the speaking person.

A way to train this is to focus on listening to the sounds around you, the conversations of people nearby, and your own internal dialogue as it responds to what you hear. Pay attention to the subtleties in your head and your choice of words. Listening is one of your most crucial senses, and along with vision, it influences your reaction to others' behaviours. Therefore, observe without judgement and stick to the facts.

RECOGNISING YOUR BELIEFS

Mahatma Gandhi had a simple model for beliefs. He said:

- Your beliefs become your thoughts.
- Your thoughts become your words.
- Your words become your actions.
- Your actions become your habits.
- Your habits become your values.
- Your values become your destiny.

Our positive and negative beliefs affect many aspects of our lives – not just our health. Thoughts and opinions control more than we realise. That's why your motivation, commitment, and focus are so important for what happens in your project and in your life. Consequently, the degree to which you achieve your desires is contingent upon your understanding of your own mental processes.

We process and interpret information based on our dominant senses and the way our brain governs our thought processes. These processes, shaped over a lifetime, form our preferences. You may have a colleague who always adopts a 'glass half empty' perspective, constantly feeling that something is lacking. On the contrary, another colleague might view 'the glass as half full' and maintain an optimistic approach to task completion.

The placebo effect is defined slightly differently but generally as 'an inactive treatment' or as 'a nonspecific, psychological or psychophysiological therapeutic effect produced by a placebo'. Bruce Lipton calls the placebo effect the belief effect (Lipton, 2015), as our perceptions and beliefs influence our behaviour and body, whether right or wrong, and thus act as a placebo.

The latest research in epigenetics – the science of how environmental factors select, modify, and regulate gene activity – seems to indicate that genes do not as we once believed determine your biological faith. But rather that genes are influenced by environmental factors such as nutrition, and various forms of emotional stress (Lipton, 2015). In this way, your focus, reaction patterns, and unconscious patterns can change the expression of genes without changing the basic blueprint. Bruce Lipton believes it is possible to use willpower to change how we react to environmental factors and thus change the epigenetic programming of our genes. This simply emphasises the importance of raising awareness of one's beliefs and any strengths or difficulties they cause.

Suppose we change the brain's frame of reference by offering it new perspectives, attitudes, and beliefs. In that case, we can also change our reactions and emotional patterns and, thus, our actions and behavioural patterns. According to Bruce Lipton, we 'download' the behaviours and beliefs we need on an ongoing basis and embed them via the synapses and neural pathways of our subconscious mind. When you learn something new, it creates new network points and new connections, and an update of your subconscious programmes – your 'model of the world' – your truth that unconsciously shapes your behaviour and opportunities throughout life. In some cases, the pre-programming of the subconscious will require us to make a conscious choice if we want to change the shape and beliefs. Lipton says that the subconscious mind should be seen as a database or hard drive that is entirely devoid of emotions. It consists of a series of programmes triggered by sensory impressions or internal signals in the form of basic emotions. These programmes dictate what we say, think, and do without taking a stance or questioning their validity.

Sometimes, we consciously disregard certain impressions because they are not exciting or don't arouse our curiosity. This can often lead to our perception of a meeting being different from that of other participants. Additionally, our subconscious programmes might dismiss information that doesn't align with our existing beliefs. Project managers who unconsciously generalise in their project work can stifle creative processes. When you encounter words like 'never' and 'always', encourage open dialogue by demonstrating curiosity and asking clarifying questions. Explore if there are any alternative possibilities.

We always interpret and distort sensory input – we want it to fit into our understanding of the truth. So, even if we receive sensory information that carries new knowledge, we may choose to perceive it differently. Each person has their own mental framework – a unique model of the world. We tend to sense other people's models of the world based on our own model of the truth. This does not produce the best results. Instead, seek to understand what lies behind the beliefs and behaviour patterns of others.

By practising self-reflection and empathy towards others, you may uncover unexpected convictions. Pay attention to feelings, thoughts, and actions – do they indicate a need for change in your belief system? Beliefs ingrained through upbringing can be the most challenging to alter. However, modifying our beliefs can profoundly impact our behavioural patterns and actions.

CONNECT TO YOUR HEART

Emotional stress reduces your brain's performance by as much as 75% and can wreak havoc on the brain's overall understanding. Under pressure, information passes through a neurological filtering system, which means that the most significant and fearful part of the information colour your reactions to the information. This inhibits your rational thinking. Therefore, in crises, the limbic brain may control the rest of the brain via the amygdala. The amygdala acts as an emotional memory bank. It stores past 'dangerous' events and dictates your reaction to situations in which you perceive a threat – even if it's not an actual threat to your life. This automatic protective mechanism triggers fear in difficult situations. Consequently, we are often overruled by our amygdala and the mental and emotional patterns established throughout fearful experiences in our lives. Because these reactions are based on your brain's subjective judgement, others may not perceive the same threat you feel. As a result, your actions may be seen as inappropriate, incorrect, or excessive for the given situation.

When a stressful event has been going on for too long, the amygdala can become over-reactive. This leads to inappropriate reactions and emotions and can cause 'tunnel vision' around the issues causing stress. Avoiding amygdala reactions is not always easy, but the better you know yourself and the situations where the amygdala is activated, the better you can control your reaction patterns.

In the often-chaotic world of project managers, amygdala, therefore, can be too easily activated. The point here is not necessarily to avoid activation but to train yourself to recognise the triggers that cause amygdala reactions. By identifying triggers, you can reduce the number of activations, choose a different path, or make sure you regain full access to your rationality before you react. This improved emotional resilience allows you to consider more options for action.

HeartMath was founded in 1991 by Doc Childre and Howard Martin. HeartMath has researched to understand the relationship between the physical heart and emotions and thoughts. A fundamental discovery is that heart rate variability (HRV) reflects how balanced our internal nerve system's state is. HRV reflects the minor variations in the time between each heartbeat. HeartMath's main product is a sensor to measure this variation in heart rate. By training focus on the heart, while analysing feedback on the heart rhythm pattern, the HeartMath system helps people to regulate their own internal nervous system. This may among other things, reduce stress, increase resilience, and generally help people to let go of anxiety and unlock the ability to make their own effective choices. HRV is currently the subject of a lot of research and seems able to reflect the balance between the sympathetic nervous system and the parasympathetic nervous system.

A high HRV generally indicates a good balance, whereas a low variation means the system is in a fight-or-flight mode. This is good when fight or flight is needed, but in most everyday situations, a balanced state is more effective.

HeartMath research shows that the heart communicates with the brain through several channels: the nervous system, hormones, neurotransmitters, pressure waves, and electromagnetic fields. In addition, the heart generates a powerful magnetic and electric field 5000 times stronger than the brain. The magnetic and electric fields can be detected 3 m from the body in all directions, affecting all bioelectromagnetic systems within its range. This suggests that a person's heart pattern can affect those nearby and vice versa. This is one of the explanations for why we can sense when other people are stressed or in good inner balance (https://www.hearthmath.org, 2023).

Research shows that imbalance and stressful situations shift the focus to the mental domain, and the connection to the heart is easily lost – both physically and metaphorically. There are simple techniques to restore brain–heart harmony, such as simple meditation techniques focusing on the heart and breathing simultaneously. From a scientific viewpoint, the complexity of the heart's nervous system is enough to consider it a small brain in its own right. Thus, the heart has its own small complex nervous system, which HeartMath calls the 'cardiac brain'. The heart sends far more information to the brain than the brain sends to the heart. The signal from the heart has a specific effect on the brain centre involving strategic thinking, reaction time, and self-regulation (Martin & Childre, 1999).

Experiencing heart intelligence with continuity requires building a reliable partnership between the heart and the head – a partnership that starts when you learn to distinguish between these two very different sources of intelligence and can tell when your thoughts and emotions are being guided by the heart and when they are not. Most people who practice techniques that connect the mind and heart – typically through meditation – find that they can regulate their emotional outbursts, build resilience, and positively shift their baseline. Sustained practice of meditation techniques will therefore increase your resilience to stress. Therefore, meditation will, in one way or another, be an integral part of the personal practice of sustainable management. Many have experienced how meditation can contribute to a kind of inner peace – and thus an inner sustainability. This is probably one of the reasons why the popularity of meditation is increasing. Meditation doesn't have to be big and fancy. During meditation, you can see how HRV improves and stabilises. Observing how small changes in thoughts and internal states cause changes directly in HRV measurements within seconds is interesting.

YOUR ENERGY

Energy follows attention, so if you as a project manager misplace your attention, the project may experience trouble. The change a project creates comes from the working point that the project manager can create and work from. If the inner working point is not well-founded and well-functioning, it will be difficult to manage the project in a good way. Therefore, it is important how you think, speak, and act as a project manager. There are numerous aspects to concentrate on, making it essential to adopt a structured approach to self-care and rational problem solving, ensuring both receive timely and appropriate attention.

We often use expressions about energy in various contexts in daily life, typically referring to the physical or human world. Over centuries, theories and methods have been developed to understand the concept of energy in the physical world. This understanding provides valuable metaphors for working with the notion of energy in our own bodies and lives. Exploring human energy can help us achieve balance in day-to-day activities. The term 'energy' implies action or work, and when discussing human energy, we refer to the vitality within our bodies and the body as an energy source. This energy stems from the food we consume and is influenced by how we manage it through our mental, emotional, and spiritual processes. Our energy levels depend on and are maintained by a combination of physical, psychological, emotional, and spiritual functions. Ultimately, it revolves around our diet, thoughts, reactions, and most importantly, our beliefs. Although we cannot directly see this energy, we can feel its presence and see what it can do.

Physical energy can create heat, light, and movement. Human energy also creates warmth, light, and movement in how we experience a person. You might be familiar with expressions like: 'I'm happy and full of energy today!!!' or 'I'm exhausted and have used up all my energy'. We utilise the term energy to describe our state; other people will also perceive our energy level in some sense. We don't measure our internal 'joule value', but you can ask yourself what your energy level is today and evaluate it on a scale of 1–10 accurately. Consider appraising your energy based on the physical, psychological, mental, and spiritual states. With practice, you'll learn to recognise bodily sensations without taking these assessments. This type of evaluation can help you identify your needs. What do you need to energise you today? All four domains consume energy. If you have been running long distances or working out for hours at the gym, you will experience fatigue, which can be attributed to your physical energy levels. But you may also recognise that you can feel tired after attending many meetings, even if you haven't moved your body. This is a sign that your mental energy is in deficit. Balancing energy in everyday life involves considering the body as a whole and ensuring sufficient energy for optimal communication between the body, senses, nervous system, brain, and heart.

In the world of physics, we talk about kinetic and potential energy, which make up total mechanical energy together. If you apply this concept to your internal energy, it is equivalent to being in flow or being at a standstill. We need both forms of energy, which we refer to as doing and being. Too much doing will create too little space for being and vice versa. As a sustainable project manager, it is necessary to work with awareness of what energises us and what drains us of energy. In this way, you can create a sustainable energy balance within yourself. We often call it 'recharging' – and that's what it really means. Recharging can be better realised if you know what type of energy you have 'run out of'.

The thought activity of the mind is a major contributor to energy expenditure. Meditation can help to create stillness by allowing the mind to rest, thus allowing the body to be recharged by going inwards. Numerous meditation techniques exist, and they don't need to consume much time.

Buddhists have known for thousands of years that meditation reduces stress and improves your overall health. Through the last 30–40 years of Western research, it is now also widely recognised in the West. Goleman and Davidson (Goleman, 2018) describe five levels of meditation use:

- Level 5: this level is even more accessible and feasible to implement in our busy lives. The level is still in its early stages of development, but could be, for example, creating moments of mindfulness while driving to work, cleaning, and so on.
- Level 4: here the meditation content is made easily accessible to everyone, for example in small meditation sessions of a few minutes. There are plenty of apps and YouTube videos that are easy to follow – even for the untrained.
- Level 3: here the spirituality is typically taken out of meditation and more of a mindfulness approach to meditation is used. Mindfulness-based stress reduction, developed by Jon Kabat-Zin, or transcendental meditation, are good examples.
- Level 2: meditation is still a central part of life, but not a full lifestyle. In the West, meditation practice is typically adapted to the Western mindset.
- Level 1: the most intensive type of practice, used by Buddhist monks, for example.

By examining meditation from this perspective, the mystery surrounding it can be demystified if needed. It's fascinating how ancient traditions and modern scientific approaches with tangible measurements converge in today's world. For instance, alterations in the brain due to meditation can be seen. Furthermore, utilising a straightforward and affordable device like the one from HeartMath allows you to directly and instantly observe how your heart reacts to meditative practices.

There are numerous approaches to meditation. The most fundamental one is to be patient and calm your mind – which is more difficult than you might think. Meditation has a quick and direct effect, which you should be able to experience the first few times you try it. But meditation over time creates a different and more lasting and immersive effect on one's well-being.

You can meditate regularly or occasionally when you need to rest. There are numerous guided meditation videos on YouTube. Explore various options to find one that matches your preferences, be it long or short, guided or silent. Remember, practice makes perfect. If (when!) an inner dialogue emerges during your meditation, simply acknowledge it and return to your focus. Meditation is about cultivating awareness, not controlling thoughts.

Meditation can be practiced while sitting, lying down, walking, or running, yielding varying results based on your physical condition. Typically, running generates more good ideas, whereas lying-down meditation is calming and healing for the body. Activities such as a walking in nature or gardening can be soothing for some individuals as well. The key is to discover the form that suits you best and find the time and space to engage in meditation. Consider and choose what you need most.

A physiological understanding of the effects of meditation can be gained through an understanding of the autonomic nervous system, which runs throughout the body. The autonomic nervous system maintains equilibrium in the body's internal state through a process called homoeostasis. The autonomic nervous system functions independently (autonomously) of consciousness, picking up information from the senses and sending instructions to muscles, organs, and glands. The autonomic nervous system has the important task of balancing the opposing effects of the sympathetic and parasympathetic nervous systems. The sympathetic system activates the body for action, and the parasympathetic system restores to normal and thus maintains energy in the body. Meditation activates the parasympathetic nervous system and relaxes the muscles, stabilising breathing and thereby lowering the heart rate and blood pressure. In addition, research has shown that we release more happiness and less stress hormones if we meditate regularly. Meditation thus contributes significantly to our inner balance. It increases our resilience to being pushed out of balance and ensures that we regain our balance more quickly when we are in a fight/flight state.

MAIN POINTS

Focusing on one's inner life involves being attentive to sensory experiences and personal preferences, which further enables the development of reading others' preferences. This helps minimise the risk of misunderstandings between individuals.

We all have our own model of the world as a frame of reference, programmed throughout our lives. This means that if we always do the same thing, we will essentially get the same result. Therefore, areas you want to strengthen or change require a focus on changing your beliefs, actions, and habits.

Emotional stress disrupts the whole brain and activates the amygdala. An over-activated amygdala means inappropriate/reactive emotions, reactions, and 'tunnel vision' about problems. The better you know yourself and the situations that trigger your amygdala, the better you can control your response patterns.

Balancing the connection between your heart and brain can enhance resilience to stress, challenges, and adversity. This balance promotes personal well-being, creativity, insight, and overall health. On the contrary, continuous energy expenditure without proper rest and recovery may result in burnout, mistakes, health issues, and diminished performance.

A sustainable project manager leverages their understanding of the body's energy systems and processes. They ensure sufficient energy is provided for the body, senses, nervous system, as well as facilitating communication between various sections of the brain and heart.

In summary, to be an effective sustainable project manager, there are a few inner settings that are important to pay attention to:

- Understanding your sensory system.
- Be aware of your own model of the world and its limitations.
- Understand your needs, interests, and wishes.
- Focus on body balance and resilience.
- Connecting heart and brain processes.

- Be ready to admit that you don't know everything.
- Recognise that our individual selves are too limited to do everything ourselves.
- Accept that life doesn't always go the way you planned.
- Many of us are limited by our beliefs and our intellectual knowledge, be open to change.

YOUR REFLECTIONS

Did anything surprise you or did your awareness sharpen after reading the chapter? Have you gained any new insights to experiment with or work on? Are you cognisant of your sensory impressions? Can you identify your personal preferences?

To make progress in the next few days, consider focusing on the filters you use and when you use them. Additionally, take note of your colleagues' language usage in relation to everyday challenges.

Reflect on the past week of your current project
Were there moments where you could have acted differently? Identify a specific situation in which you felt you fell short, such as a conversation or meeting. Spend a few minutes reflecting on this situation. Did your positive intentions take centre stage, or did negative expectations dominate? Were alternative options available? Contemplate whether you should choose a different approach if faced with a similar situation in the future.

In what ways can you become more aware of your beliefs?
Understanding your beliefs requires focus and energy and awareness of the motives behind your thoughts, words, and actions. Go backwards into your beliefs via your actions and check it out. Do they fit how you see yourself or how you want to see yourself? Where do your strengths lie?

Are you a good listener? Are you aware of what is happening around you?
Active listening is a concept focused on mindful thinking. By practising active listening, you engage in being present and truly understanding the other person's perspective. How do you hone your listening skills? How do you absorb new information?

How do you experience stressful situations?
We all experience situations of short-term stress due to our behavioural patterns. Are you aware of your 'stress programmes' – do you find yourself overreacting in some situations? Take a few minutes to think about your last 'disaster situation' in the project. Can you recall the situation? And can you recall your thoughts and your reaction? Can you feel your trigger points? To what extent was your reaction appropriate and how much was it an overreaction?

How strong is your connection to your heart?
The heart continuously communicates with the brain through the exchange of electromagnetic fields, influencing those around you. Have you had experiences

that support this research? How powerful is your heart? Do you possess high resilience in receiving emotional information from your environment?

If you want to start strengthening your connection to your heart right now, you can do this exercise:

(1) Focus your attention in the area around your heart.
(2) Imagine that your breath is going in and out through the heart area more slowly than usual (5 s inhalation, 5 s exhalation, or maybe longer).
(3) Continue for 5–10 min or until you feel a connection to your heart.

Or

(1) Make a sincere attempt to experience a regenerative feeling like appreciation or care for someone or something in your life.
(2) Try to rekindle the feeling for someone you really care about (children, pets, favourite place, etc.).
(3) Feel it – how can you feel it in your body? In your heart?

Can you feel your own energy?

Measuring your energy level is an internal state where you ask yourself what your energy level is today on a scale of 1–10. Do you have an awareness of your own energy? Can you distinguish the energy of the four intelligences (physical, mental, emotional, and spiritual)?

How and how often do you charge?

Where and how do you recharge? Sit down and relax your body and take a few deep breaths to relax your body. You can focus on your breathing or on one of your senses. You will find that your mind will start to relax too. A little trick is to imagine that your thoughts are little clouds in the sky being carried by the wind. Then imagine that thoughts that are not important are put on the clouds and drift out of your inner field of vision. After a few times, you will find that unimportant thoughts disappear by themselves on the clouds in the sky. Remain seated until you feel the energy returning.

READ MORE

If you want to know more, you can find more inspiration here:

'The Intelligence of Emotions' by Daniel Goleman explains how our emotions and the way we regulate and read them affect our lives (Goleman, 2005). The topic is further elaborated on and made applicable to leadership in 'Primal Leadership' by Daniel Goleman, Richard Boyatzis, and Annie Maackee (Daniel Goleman, 2016).

'You are the Placebo – Making Your Mind Matter' by chiropractor Joe Dispenza talks about the placebo effect and how thoughts and beliefs have a significant influence on the body's functions (Dispenza, Your are the Placebo – Making Your Mind Matter, 2014).

HeartMath has published many books on research into the effects of the heart on us. For example, read 'The HeartMath Solution' by Doc Lew Childre,

Howard Martin (authors), and Donna Beech, which describes the studies and measurements made on HeartMath (Childre & Martin, 1999).

'Your Intelligent Cells, the Biology of Beliefs: The Miraculous Powers of Mind and Matter' by Bruce Lipton on, among other things, epigenetic effects and how our beliefs guide us (Lipton, 2015).

Eckhart Tolle has written two books, 'The Power of Now' and 'The New Earth'. 'The Power of Now' is about being present and letting go of our worries about the future (Tolle, The Power of Now: A Guide to Spiritual Enlightenment, 2004). 'A New Earth' is about shifting our focus to the present and accepting the moment – making a shift in consciousness (Tolle, A New Earth: Awakening to Your Life's Purpose, 2008).

The book 'First Intelligence – Use of Science and Spiritual Intuition' describes how you can work on developing your intuition (Wright, 2014).

'Our Conscious Universe' by Eben Alexander and Karen Newell focuses on the consciousness paradigm shift we are looking into globally and that we are all part of a shared collective consciousness; a book about the intersection of spirituality and science (Alexander, 2017).

If you want to read more about meditation and what meditation can do to your body, mind, and emotions, the book 'The Science of Meditation' is a good description (Goleman, 2018). Or read the book 'Becoming Supernatural' by Joe Dispenza, which describes the latest research in neuroscience, epigenetics, and quantum physics in the context of meditation training (Dispenza, Becoming Supernatural, 2019).

doi: 10.2166/9781789063349_0159

Chapter 6
Being self-disciplined

There is no magic wand that can solve our problems. The solution lies in our work and discipline.

Jose Eduardo dos Santos

There are numerous subtle ways in which we evolve without realising it: embracing new technologies, encountering inspiring individuals, and consuming news and films that influence our perspectives on the world. This chapter focuses on consciously and intentionally transforming ourselves. It delves into harnessing our willpower and actively training it for deliberate change.

To achieve sustainable project management, substantial willpower is essential. This willpower can be honed and developed through consistent practice. It's your determination that brings ideals to life. Begin with your current willpower level and progressively adapt your actions to align with your reliable capabilities. Your principles, ideals, and values will guide your growth. Development involves aligning your thoughts, speech, and actions with your designated ideals more closely over time.

This chapter outlines various approaches to change. Once you've identified the area you wish to change, you can follow a straightforward model in a disciplined manner. When working on self-improvement, we typically employ a strategy of small, daily, and consistent changes. This approach helps avoid feeling overwhelmed while boosting your self-confidence and cultivating patience. Although we have experimented with bold transformations, experience shows again and again that awareness and modest adjustments yield the highest success rates and most significant impact in our lives. Numerous small actions can accumulate to substantial results.

> The ability to observe without judgement is the highest form of intelligence.
> Krishnamurti (Indian philosopher)

Authors: Pernille Ingildsen and Mette Thorkilsen

INTENTION AND DIRECTION

> 'Please tell me which way I should go from here?' 'That Depends very much on where you want to go', said the cat 'I don't really care', said Alice. 'Then it doesn't matter which way you go', said the Cat. Alice in Wonderland

Daily life in projects can be quite tumultuous. One moment, a rational approach is required to solve a problem, and the next, emotional intelligence is needed to collaborate or engage in dialogue with stakeholders. We must strike a balance between rationality and emotion, using both thinking and feeling to determine necessary adjustments. Taking responsibility for managing and responding to our emotions is crucial, as we are the only ones who can influence them through our inner attitudes and thoughts.

As a project manager striving for a sustainability in our projects, it's essential to be mindful of thoughts and assess whether they accurately represent reality. We often place too much trust in our internal monologue, whose main task generally focuses on keeping us safe. This inner voice tends to rewrite our history, portraying us as 'the good guys' and external challenges as 'the bad guys'. However, this perspective is not conducive to effective project management. Instead, adopting a more realistic view of the situation enables us to take better action.

In leadership, you need to be generous in your interpretation of others. If you do not understand other people's actions, the task is not to invent stories of evil, incompetence, deceit, and so on. The task is to understand why they are doing as they are. You can be confident that the action you are recording looks rational and aligned in the eyes of the person carrying it out. Moreover, they likely perceive themselves as one of the 'good guys'. So, if an action is inappropriate for the project, the task is: 'to understand'. If you allow your mental dialogue to spin a story about what is wrong, you are in effect closing the connection between you and the other more and more. Try not to see yourself as a victim of the other person's action. Rather, see yourself as the protagonist who investigates and corrects. And beware that sometimes, it is yourself who needs correction.

A helpful quote is provided by Dean Inge (as quoted by Robert Greenleaf in Servant Leadership) (Greenleaf, Servant Leadership, 2002): 'Faith is the choice of the nobler hypothesis'. When encountering troublesome behaviour, take time to brainstorm possible causes, and opt for the most realistic and noble hypothesis about your counterpart. With this hypothesis as a foundation, collaboratively seek solutions. View yourself as a co-creator of actions, behaviours, and thought patterns alongside those involved in the project.

Strive to become more aware of your communication tone. Often, the initial information sets the meeting's tone, theme, and overall atmosphere. Be cautious not to let negative expectations about the meeting, participants, or topic becomes a self-fulfilling prophecy. Poor communication can have significant costs in both outcomes and personal relationships. Remain conscious of how your communication shapes situations and consider adjusting your attitudes and values as necessary.

To achieve success, it's often necessary to take the counterintuitive step of moving closer to those you'd rather distance yourself from. This moving-closer demands courage and energy to overcome personal resistance. If you do not address a challenging situation, it will have a tendency to grow more difficult as long as you ignore or avoid it.

Only when you start to challenge your beliefs will you be able to change your own communication. Beliefs give stability to our everyday life. Awareness of your guiding beliefs is the starting point for balancing your communication and behaviour. Try to challenge your own understanding of the truth – often there is also a grain of truth in the opposite of what you believe.

We all have blind spots where we may be unaware of the consequences of our actions, while others can observe these effects, both positively and negatively. Our perceptions of each other differ because we identify with different aspects of ourselves in others. Our shadow sides often become intertwined with our relationships, which is why examining them is crucial when seeking to improve our connections. These shadow sides manifest when we emotionally react to other people's actions with contempt, judgement, irritation, or anger. Essentially, what we dislike in others often indicates our own shadow sides.

Although confronting these aspects may be uncomfortable, it's essential to recognise that they also hold positive attributes and reveal untapped potential. Accepting and using them constructively is key. An indication that you've addressed your shadow sides is your ability to laugh at yourself and not be easily offended. Embracing your shadow sides allows you to better recognise, understand, and accept differing perspectives from others. This skill helps you better identify yourself in others – also when they fall short.

So, what essential changes will help you grow as a better leader? Which behaviours, thought patterns, or skills do you want to improve within yourself? Your starting point and ultimate goal will determine this.

In this chapter, we introduce various methods to help you define your ideal goal and take control of your actions to implement desired changes. Primarily, change involves consistently practising new behaviours. All change takes time, so patience with our own development is crucial. Eventually, the new approach will become a natural part of your identity and actions.

ABOUT BEING PROACTIVE

Inspiration can be drawn from Stephen Covey, who, in 1989, published one of the best-selling books on self-management (Covey, 2020). This book presents seven habits or principles that guide individuals in creating successful experiences both personally and professionally. According to Covey, the development of successful habits hinges upon willpower, knowledge, and skills. We must possess the determination to change our behaviours and habits, understand the methods to do so, and have the physical, emotional, and mental capabilities to execute these changes.

Acquiring the knowledge for change is relatively simple, as you can draw inspiration from books, colleagues, or online resources. The skill part is to creatively develop practices that allow you to build new habits while keeping realistic expectations (the SMART rule applies here). Willpower is more nuanced. It involves recognising the need for change and committing to it. Now is now. Ultimately, the decision to change comes from within, and only you can make that choice.

For example, if you want to change a habit, this is the kind of thinking that is needed:

> I've decided to adopt a less confrontational approach. I believe that by speaking less and listening more during status meetings, I'll generate less conflict. As a project manager, it's not necessary for me to dominate conversations or dictate tasks to others. Instead, I can utilize a more facilitative method. To achieve this, I'll need to develop my knowledge and skills in facilitation. Embracing this change would not only benefit me personally but also enhance my reputation as an effective and respected project manager. As such, it's crucial for me to find the motivation to transform this aspect of myself.

The first three habits in Covey's book emphasise personal development, whereas the subsequent three focus on effective interaction with others and achieving mutual success. In this chapter, we will concentrate on habits 1–3 as a source of inspiration for goal setting and acting in areas you wish to improve.

The first three habits are:

(1) Habit 1 – Be proactive.
(2) Habit 2 – Begin with the end in mind.
(3) Habit 3 – Put the first thing first.

Habit 1: Be proactive

The first habit revolves around being proactive and utilising your innate ability to make conscious choices. You bear a major responsibility for your own life, the decisions you make, and the resulting consequences. Although seemingly obvious, it's quite rare for individuals to consciously acknowledge the responsibility for their decisions. Reflect on your own life – when did you last make a significant choice? Did you consider its repercussions and were they aligned with your personal values? Did your decision enhance your capabilities or well-being? Often, choices arise as reactions to others' actions rather than stemming from a genuine aspiration for self-improvement. Independent proactive choices contribute to becoming the person you aspire to be.

You should focus on tasks within your scope of control and influence. Unfortunately, many people spend too much time on things they cannot control, often spending their time complaining – which as it turns out significantly reduces their impact. Being proactive involves choosing something within your responsibilities and possibilities – a decision to deviate from your usual actions.

It's crucial to select a goal that truly motivates and energises you. The desired outcome should be something you genuinely desire, also if it means occasionally stepping out of your comfort zone. Believing in the achievability of your goal is important, as well as visualising and sensing it. In this journey, pay attention to your senses, physiology, nutrition, strength, and endurance.

Habit 2: Begin with the end in mind

Habit 2, 'Begin with the end in mind', emphasises the importance of setting clear goals and envisioning your desired outcome. This habit incorporates self-awareness and identity, because all changes of your paradigms should align with your ultimate objectives. Just as with project tools, establishing a purpose and success criteria for altering a specific paradigm is crucial. Envision how your behaviour will transform after focusing on your chosen area of improvement. Self-awareness, imagination, and visualisation are essential when working with inner goals. A paradigm shift may also impact other beliefs and alter your identity in the process. Imagining what others would say about you in the end situation can inspire change. For instance, a goal might be: 'I want my colleagues to recognise my ability to manage meetings effectively, achieving results quickly instead of engaging in endless debates'.

Visualisation can be employed to establish personal goals and generate a vivid image of a brighter future. For instance, project participants can form a positive impression of the project's successful completion through visualisation, treating that image as a goal itself. In truth, everyone unconsciously visualises by crafting mental images of desired outcomes or situations. Vision is derived from the Latin word 'visus', which means 'seen' or 'sight', visualisation can elicit physical and emotional reactions, such as salivation, happiness, or heightened anticipation.

Whether unconsciously or consciously, your thoughts and beliefs can significantly influence your body's processes. Becoming more aware of your thoughts and intentionally directing them towards positivity can hasten the process of change. Negativity in your subconscious mind may yield unfavourable outcomes. However, if you consciously cultivate a positive mindset, your entire body will respond accordingly, leading to different behaviours, paths, and ultimately, better results.

Visualisation is a potent tool that resides within you. To harness its power, simply activate your ability to create mental images and scenes of your desires. As you generate these inner visions, observe how they ultimately materialise in reality.

The initial step can be the most challenging for various reasons. For instance, if you're occupied, it's tempting to continue working with familiar methods. You might also have a subconscious or conscious aversion to change, and uncertainties may arise regarding how the alteration will impact others' perception of you. To achieve your goals, it's essential to decline certain opportunities to focus on the critical paradigm shift. This means putting 'first things first' – a concept highlighted in habit 3.

Habit 3: Put the first thing first

> The things that matter most must never be slaves to the things that matter least' Goethe

When willpower is present, you can usually overcome initial hurdles quickly. However, in some cases, things may not progress as smoothly as anticipated. Maybe you don't fully believe in what you're doing? The sooner you take

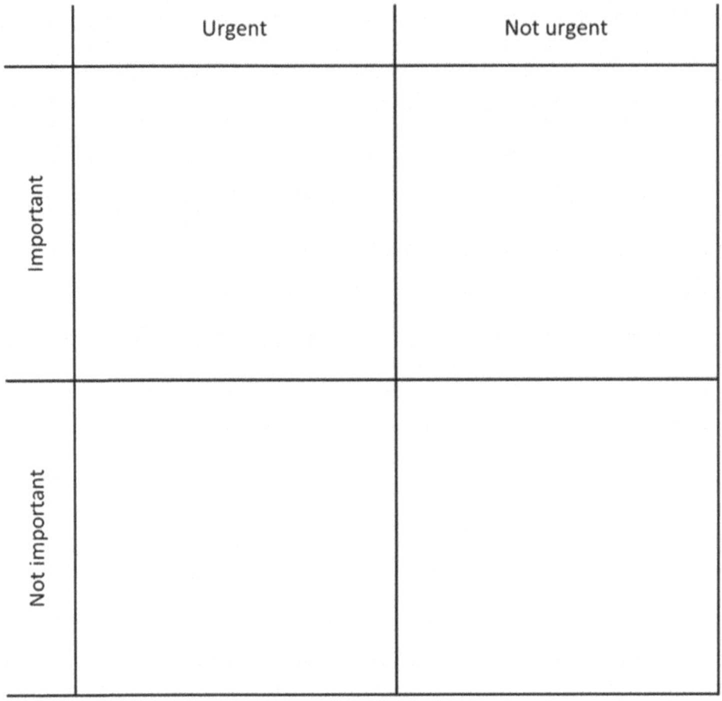

Figure 21 Stephen Covey's time matrix (Covey, 2020), which is a tool that can be used to map your own daily activities and from there prioritise what is most important in relation to your own change.

responsibility for your work, the quicker you'll see results. Similarly, as soon as you take yourself seriously, it becomes easier to prioritise and accomplish tasks efficiently.

To effectively implement change, clear some time in your calendar and stay focused on your goals; otherwise, success may remain out of reach. Covey recommends using the prioritisation matrix tool to establish a balanced daily routine shown in Figure 21. The matrix consists of two axes: 'urgent vs. non-urgent' and 'important vs. non-important creating 4 different fields'. By categorising your daily tasks, you can gain insights into your current life choices and make necessary adjustments.

If you're not careful, your time may be consumed by the urgent and important field. Is this truly wise? Covey emphasises the need to allocate time for important but non-urgent matters, acknowledging that some things take time. Continually focusing on urgent firefighting activities could prevent you from addressing what truly matters in your personal growth. The non-important field, combined with 'urgent' and 'non-urgent', may hinder your creativity. Periodically review your priorities and ensure you allocate time for both important and non-urgent tasks.

TRAIN SELF-CONFIDENCE AND PATIENCE

For many of us, rebuilding self-confidence is a crucial endeavour. The self-confidence we discuss stems from an inner ability to rely on oneself, rather than a superficial facade. It is not like an attractive paint that can be applied to the surface of a person to make them appear more confident. Instead, there is self-confidence that comes from within. It arises from consistently making decisions and changes we believe to be right, and honouring the promises we make to ourselves. As we take ourselves seriously without being overly stern, true self-confidence emerges on its own. This genuine confidence is not only visible on the outside but, more importantly, is felt on the inside. It's the result of an ongoing internal process of mindful adjustments in alignment with our ideals.

When we fail to keep promises to ourselves, we lose confidence in ourselves – like we would lose confidence in others who compromise on their promises. Every time, we don't hold up on what we promise ourselves, it is time to reflect. What was the cause? Why did it happen? Then we can correct our personal strategy going forward. Taking this work seriously is key to future sustainability. You should take your own development journey and your goals for yourself as seriously as the goals set in and by the outside world – in fact, you should probably take them even more seriously. Inner discipline is important and requires dedication, passion, confidence, and energy.

The issue arises when we choose the easy way out, ultimately disappointing ourselves. This can happen effortlessly since there isn't anyone to hold us accountable. Or is there? Indeed, there is – us! We must keep ourselves accountable for our actions. If we fail to live up to our own promises, we risk losing confidence – self-confidence.

If we constantly tell ourselves to exercise more but fail to do so, our confidence declines. The same happens when we pledge to lose weight yet continue indulging in cake, commit to meditation, vow not to scold others, or promise to be better people without taking action – our self-confidence gradually dwindles.

The solution is not to push harder, overstep inner boundaries, adopt more of a 'shape up or ship out' attitude, starve us to death, or set marathon-like ambitions when we have just started running. The best solution is the opposite – to reduce our ambitions. We need to promise ourselves changes that are so small that we are 100% sure we can keep them. When we make promises to ourselves and keep them over and over and over again, over time, we gradually build up the confidence we have lost.

Confidence comes from:

(1) Practice reflection before making decisions to change, rather than impulsively choosing New Year's resolutions in the spur of a moment.
(2) When you identify something as important, act. True confidence stems from trusting the promises you make to yourself. Therefore, be mindful of the commitments you make – they should be both significant and achievable.

This inner confidence will naturally radiate through your outward demeanour. Such confidence will also manifest in your relationships, allowing others to depend on you and effortlessly follow your lead.

People trust you as a project manager because they see someone capable of self-leadership, and therefore, able to lead others. It's true that we can only effectively lead others when we master leading ourselves. Self-leadership isn't about appearances; it's far more challenging than simply looking like a leader: after all, if you've got a few stars on your shoulder and start throwing orders around, surely, you're leading others? Although it's easy to give orders and seemingly take charge, true leadership begins with understanding and guiding oneself through change. This involves knowing what transformations you're undergoing and why, as well as ensuring that you remain focused until the process is successfully completed.

Therefore, start experimenting with changing yourself in small ways. Focus on understanding your own change process; build self-confidence and self-discipline by finding a process that works for you.

A vague goal to 'get better' won't yield tangible results. Change requires clear direction, dedication, and patience. Transforming our inner mindset takes time, just as one cannot go from being short-haired to long-haired in a week. It takes time, which is why impatience is mostly a problem and rarely a strength. When some people these days half-proudly talk about how impatient they are and try to show how over-efficient and 'pushy' they are, it's the wrong way around. Patience is the ultimate currency, in relation to yourself, in relation to others, and in relation to the whole. Impatience creates pressure and rash decisions. Patience is not the same as working (too) slowly. Instead, it is about influencing the systems you work in with the right amount and direction of pressure at the right time. Forcing matters rarely leads to positive outcomes in the long run, and presently, you're seeking the assurance of favourable results over time.

In your training and change work, pay attention to sharpening your sensory inputs so that you know how you and others react to your actions. You may notice things you have never thought of before – the road will often surprise you. You may fail the first few times you choose to do something different from what you usually do, but this becomes less important if you have a high degree of flexibility. If you find that your choice wasn't the right one or didn't work out, try with another option – try something else.

That is why it is so important to set-up a process around yourself that facilitates you to keep a conscious focus and consistent effort while the 'organic change process' takes place. In that way you guide the process within you instead of forcing it. A slower process is often safer than the one that is fast. Numerous factors can impede sustainable project management, including mental barriers, relationship issues, or structural obstacles. However, these barriers can typically be overcome with patience and a series of small, conscious changes over time.

Despite the amusement it may bring, we hold immense respect for those who repeatedly take stop-smoking courses. Each attempt to achieve a goal demands effort from us; every time we strive for self-improvement, we gain new insights about ourselves. But it is also a frustrating process to keep trying and failing. This is why connecting reflection and theory to these efforts is advantageous. That's why unsuccessful attempts should be valued like gold. What lessons have I learned? How can I increase my chances of success in the future? How should I adjust my next endeavour?

LEAST POSSIBLE CHANGES

'Least possible change' is a method that focuses on implementing numerous small changes systematically and effectively to contribute to better self-management (Hildebrandt, 2016). The method encourages embracing a new direction in one's life and, step-by-step, discovering what works based on faith in one's ability to gain deeper understanding throughout the process.

During the process, you explore with a sense of curiosity what works for you and what doesn't. When you learn to adapt yourself in an open-minded manner, you gain a deeper understanding of how to change your own system as well as the systems you're a part of. In essence, by leading yourself, you gain experience in fostering change within a system by understanding yourself as a complex system. The method has two goals: to succeed in creating change overall, meaning refining an open change process, and to achieve the specific changes you desire for yourself.

Over time, the goal is to establish a transformative process that yields positive results for you. Along the way it's important to practice self-awareness, recognising what does and doesn't work for you when you change your system as you modify your approach. According to the sensory-based method, a key aspect involves cultivating a deeper connection with yourself, enhancing your ability to interpret your system's intuitive recommendations for adjustments. Intuition encompasses your rationality, emotions, bodily awareness, and relationships. Therefore, honing our intuition involves tuning into our inner selves and receiving guidance from our internal orientation systems.

In this model, numerous small actions accumulate to yield significant results. These many actions pave the way for change, keeping the process engaging, challenging, and 'alive'.

Method

The foremost principle to understand is that adopting a slow approach is more beneficial than moving too quickly. Hence, this transformation process was termed 'least possible change' by Stubberup and Hildebrandt (Hildebrandt, 2016). This term underscores the importance of implementing small changes for these adjustments. Smaller changes are more effective because it is easier and therefore easier to be consistent.

The method involves making continuous small changes by committing to one change at the time and then fulfilling that commitment to oneself. Then the process is repeated – commit to a new change that makes sense. Do this at a comfortable pace, allowing change to accumulate through the effects of these incremental adjustments. Experimentation stems from reflecting on the changes and the overall process. The approach encourages us to feel the impact of one move of change within ourselves before committing to a new. Although the process may appear simple, it has the potential for achieving a high level of sophistication in its practice.

The process does not require a strict timetable; however, for us maintaining momentum calls for a structured approach. In the absence of a well-defined system, the intentions to change practice often falter.

The two steps

The centrepiece of the process is the progression of small changes. A principle for selecting a change is that it should be minor and innocuous enough to not trigger your fear response (amygdala). Avoid burning platforms, drastic measures, grandiose actions, or sweeping day-to-day visions. From day-to-day, it's just small steps, one foot in front of the other in the direction you've decided is meaningful for you. A good starting point for most individuals is focusing on self-care. Subsequently, one can work on enhancing leadership skills, such as improving listening abilities. For instance, choosing a change aligned with one of the sustainable development goals can be very fulfilling and you may surprise yourself of what you are capable of. People often underestimate themselves and the impact they can make through focused and disciplined efforts. Discussing 'what should be done' and expressing scepticism doesn't solve problems. Although talking about issues may give the illusion of contribution, it is through taking action that something real happens.

Change happens when you promise yourself to carry out an action (a minimum change) and do it – and then you repeat the process.

It doesn't matter how you decide on your least possible change.

Each of the small actions should be so small and manageable that you can confidently complete the activity. For instance, rather than five larger actions in a month, a steady stream of 30 small actions in a month will often work better. In principle, the 30 small actions could form the puzzle pieces of an overall project, but they don't have to. If you take your biology seriously in this challenge, the point is to take in feedback and practice getting better at sensing and deciding what tomorrow's action should be each day. That is, to look at your context as it is on day x and decide on day $x + 1$ – and having no further plans ready.

We suggest you try to start by working on a daily change. For example, on the evening of day 1, you choose your change or action for day 2. Then, on day 2, you implement it, and in the evening, you decide what your change should be for day 3, and so on.

To enhance your ability to accurately listen to your internal feedback system, it's helpful to assess your emotions each evening before deciding on any changes. One effective approach is to review the day's events and determine the two most positive and most negative occurrences. For both the positive and negative moments take a short time to sense again how that felt and make a note on it. Feel into how your nervous system processes the events of the day. By briefly reliving the emotions associated with each situation, you can fine-tune your awareness and better understand your responses.

You can write down the two most positive and negative events to help you focus on your inner feelings. Some people don't like the words 'most positive/negative' but prefer alternative terms, such as 'what works/doesn't work', 'balanced/unbalanced', 'feels right/wrong', 'good/bad', or 'energising/draining'.

Explore your inner self to discover the terminology that resonates best with you and utilise it. This exercise is about tailoring the process to your needs, focusing solely on what works best for you – and nothing else. Once you have spent time connecting to your own feedback system, it is often more apparent for you what tomorrow's change should be.

First, make sure you specify the action explicitly so that you will be able to accurately determine afterwards whether you have completed the action or not. Second, make sure it is small enough that you are confident you can commit to it. Remember that every time you make a promise to yourself and don't keep it, you make a withdrawal on your confidence account. Similarly, every time you keep your promises to yourself, you add to the account – every day counts. Third, ensure that the action depends only on your efforts and not on others doing this or that – otherwise you end in some quagmire of your own making.

Changes should be brief and only demand a small effort, allowing seamless integration with your existing commitments. Avoid turning it into a kind of 'performance optimisation' and focus on inner motivation. Heed your inner voice; pursue actions that align with your true nature or highest aspiration.

At the end of each day, when it's time to go to bed, it's also an opportunity to reflect on whether you succeeded in completing the small action of the day. If the answer is no, then rushing to do it at the last minute is not a good idea. Instead, let it pass. Note to yourself whether you succeeded – maybe you only managed to do it 50%? At the end of the month, you can calculate the average for all the days and figure out how much confidence you can have in yourself. From a recent experiment, we found to be happy with 60% confidence in our ability to do these daily tasks, and for us it's uncommon for a month to surpass 85% – despite the ultimate goal being 100%. Regardless of the average result, understanding your level of self-confidence is beneficial. In fact, maintaining 85% self-confidence is quite impressive.

The reflection process

Overlaying the two-step process there should be a reflection phase. To simplify, we can envision the two-step cycle occurring daily and the overlying reflection process typically taking place every 2 weeks or once a month.

The instruction for the overlying reflection process is to take a 'scientist-like' look at your process. Look at the two most positive and the two most negative emotions, the action, and the degree of success. Some people may have completely forgotten that they were doing the process – it happens. But regardless, practice not moralising or judging your outcome – it's not conducive. Instead, take care of yourself and enjoy what you have achieved. Regardless, it's more than if you had done nothing.

Reflect on what worked best for you, as well as the obstacles you've encountered. When reflecting, examine your achievements objectively, as if they were someone else's. The key is to contemplate how you operate, what drives you, what hinders you, and the rationale behind the goals you've set.

The key is to lead yourself. Learning involves setting goals, acting, and following up. So, after the first run-through and reflection process, you might want to adjust the process. Maybe the notes should be made on your phone instead of on a piece of paper, maybe before you go to bed, maybe in bed, maybe in the morning. Maybe you need to set a reminder to reread and thus remember tomorrow's goals and actions. When you don't succeed, what were the reasons? Fatigue, forgetfulness, or setting to high goals? You can always reduce your 'least possible changes'. Over time, you will find that you get into a 'better frame of mind' and can take on somewhat bigger goals. But you will only get there if you manage to get into the flow through very small changes. If the steps are too big, you may still be facing the first step at the end of the month.

When the change is realised

Self-leadership is about consciously guiding oneself with awareness. Awareness of the process is essential for truly getting to know yourself, as intimately as mastering a musical instrument.

Assess whether you have finished the process you started. If your motivation for the direction is gone, stop the process and start a process with a direction that you think is more important now. If this means that you often change direction, reflect on why it is so. Is it the right strategy for you? Similarly, if you stick to the same direction for a very long time, is it right for you, or has it become a habit? If it's just a habit, perhaps it's time to explore new areas for growth. Monitor your workload to ensure a balance between doing too much and too little and develop strategies to adjust for future progress.

This method trains you to focus on long-term change rather than seeking a quick fix. Lasting transformation cannot be achieved rapidly.

If you are a person who wants to make a difference, the first stumbling block is often that you have both too high and too low expectations of what you can do. We mistakenly believe (1) that we can create rapid change, and (2) that our direct actions are the only cause of change, and (3) that we will always clearly see the results of our work.

The root cause of all three issues stems from approaching the world with a fragmented perspective. By slightly adjusting these notions, we can achieve more than we ever imagined. This adaptation requires recognising our biology, encompassing not only our own physical systems, such as arms, legs, lungs, heart, brain, and nervous system but also acknowledging that societies at every level are biological systems. Our projects, departments, organisations, countries, global humanity, and the comprehensive biological network of all living beings are interconnected – they are not merely machines. So, we need to change our mindset from a machine mindset to a biology mindset. With biology it is always slow adaptation and no sudden moves.

MAIN POINTS

As a sustainable project manager, it is crucial to develop your discipline in making proactive decisions. Ensure that your actions consistently align with your strengths and capabilities.

It's easier to continue with familiar methods, but taking the next step is crucial and challenging. To become a sustainable project manager, you must establish goals, make conscious choices, and let go of outdated beliefs. Embracing uncertainty about your destination is natural, as is concern about others' perceptions when adopting a new approach. Recognise this as part of the change process. Develop a strategy for transitioning, trust in your decision, and allow for time, space, and patience.

Your senses will aid in recognising options and making choices. If you choose incorrectly at times, remain flexible and opt for a different path next time. Developing discipline involves going through the change loop repeatedly, becoming increasingly aware of your internal shifting emotions as well as external changes. Part of this attentiveness includes being present in conversations, actively listening, and perceiving subtleties in what is said. This allows you to sense the energy, emotions, word choices, and imagery used.

Utilise your personal projects as learning opportunities. Once you've determined your desired direction, apply the method of least possible change – a systematic and disciplined approach to implementing series of small adjustments that cumulatively result in substantial progression towards, for example, improved project management. A crucial success criterion is fostering self-confidence. Confidence stems from actively making your own choices, acting upon those decisions, and honouring the commitments you make to yourself.

YOUR REFLECTIONS

To successfully achieve a sustainable transition, we must actively and independently enhance our understanding, vocabulary, and mental frameworks. If we are unable to process new information on, for example, sustainability, we need to pause and reflect on why we're struggling to accept the information. It's astonishing how much discipline and effort are required to pause and reflect. We often feel tempted to bypass this process. The enjoyable sense of immersion we experience while reading a book or binge-watching Netflix series can be difficult to interrupt. Therefore, a vital skill in sustainable project management is regularly practising the art of pausing for reflection.

The previous reflection sections allowed you to hone your skills in monitoring your inner thought process. As you revisit the reflections from Chapters 4 and 5, have you pinpointed specific aspects of this process that you'd like to carry forward?

If you skipped the reflection exercises, what was your reason? It is time to work on methods that will enable you to implement the change in a disciplined and structured way. Are you ready? Are you willing? Or only half-way?

If you seek to accomplish a goal that has taken shape but has not yet become a reality, consider reflecting on the following questions, which serve as an exercise in self-awareness.

(1) What do you want? Describe it positively, and as detailed and as based on sensory input as possible.
(2) What will it give you?
(3) How do you know you have achieved what you want?
(4) Where, when, and with whom do you want it?
(5) How will it affect other areas (and people) in your life?
(6) What stops you from already having it?
(7) What resources do you already have that can help you achieve your goal?
(8) What additional resources do you need to achieve it?

(9) How will you get there? First step. Be precise, it should be easy to achieve the first step and get off to a good start, break it down into small chunks and recognise if there are multiple paths to the goal.

Reflect on your daily life and identify your top five to eight activities.

How balanced is your daily routine? Are you utilising your time effectively? If given the opportunity, what aspects would you alter? Determine what needs to be modified or relinquished for you to embrace change in a disciplined manner.

Are you aware of potential obstacles that could hinder your success in implementing changes? Personal blind spots may impede your progress. Consider two individuals you like and two you dislike. What qualities do these people possess that either attract or repel you? How does these qualities reflect yourself? Do any patterns emerge that obstruct your efforts to improve self-leadership?

Create a personalised change strategy. Which principles resonate with you? Design a change process tailored to your needs. Select a direction and begin experimenting. Document the journey to maintain momentum. Engage your senses actively and employ constructive reflection on the outcomes. Utilise the 'minimal necessary change' approach.

READ MORE

'Sustainable Leadership – Leadership with Heart' by Steen Hildebrandt and Michael Stubberup describes a new sustainable leadership mindset and new skills and competencies needed to co-create a sustainable world (Hildebrandt, 2016). Here, you can also read more about the 'least possible change' methodology.

'Indirect Work. A Regenerative Change Theory for Business, Communities, Institutions, and Humans' by Carol Sanford with Ben Haggard (Sanford, 2022). The book is about how you can create a lasting change in your life by understanding regenerative change processes. Sanford believes we can achieve transformation with a paradigm of systemic thinking. The book provides an understanding of how to develop your consciousness and engage as a systemic change agent. It describes methods for breaking the old patterns and creating a new mind that works intuitively in the context of the living systems that function.

The book 'The 7 Habits of Highly Effective People' by Stephen R. Covey (2020) is a guide to working on yourself and becoming more effective in your daily life. Covey guides you through seven areas relevant to inner leadership for co-creation with others.

'Whole Brain Living: The Anatomy of Choice and the Four Characters That Drive Our Life' (Taylor, 2022) is a book about understanding the connection between our thoughts and our emotions so that we no longer must be bound by our emotional reactions. The book is about being aware of the connection and choosing to live to be the best version of ourselves and control what's going on in our heads.

The book 'The Fifth Discipline' (Senge, 1999) describes how organisations can adopt learning strategies to improve their productivity and success. The basic idea is that new and expansive thinking patterns can be nurtured, collective aspiration is unleashed, and you learn to create the results you want deep down.

Part 3

Collaboration Shapes the Future

doi: 10.2166/9781789063349_0177

Chapter 7

Team awareness

We have to consciously study how to be tender with each other until it becomes habit.

Audre Lorde

The idea that two plus two can equal five comes from systems theory, where some systems yield capacities that are greater than expected from the sum of the resources or capacity of the parts of the systems – this can also be the case in projects. However, in projects with teams that are merely functional, and two plus two equals four. Or even worse, two plus two equals three, and then we talk about poorly functioning team cooperation. Therefore, it's worth considering which category your project group falls into. How can you make your team function in a way where two plus two equals five? It is wise for project managers to continually assess their team's performance and identify group dynamic issues requiring attention.

As project managers don't have direct authority over the team it is even more important than normal that he/she understands the essentials of leadership and team building. Often project teams consist of members from various departments with different references, which can be challenging. In the construction industry, project delivery teams may comprise parties from multiple organisations, each with their distinctive objectives, culture, and project behaviour. This certainly does not lessen the challenge of team management for the project manager. However, complex structures are a condition in the building and construction industry because so many different competencies and resources must be brought into play to succeed.

In this chapter, we introduce various tools for fostering a high-performing team that achieves great results collectively. These well-established models cover team roles, team development processes, trust-building, social dynamics, and effective conflict resolution. Each tool assists project managers in promoting team awareness, emphasising that everyone is working together towards a

Authors: Pernille Ingildsen and Mette Thorkilsen

common goal. Essentially, project managers must be versatile and creative to maximise team collaboration and fully harness their team's potential.

In addition to the tools, we believe that the project manager should work on finding the talent in the individual, cultivate a trust-based leadership style, and create a respectful and caring tone in the team. No matter how much energy you put into developing your team, there's no guarantee that conflicts won't arise along the way. However, maintaining a good team spirit greatly impacts the topics and course of conflicts. Team awareness is a moving target that needs to be nurtured on an ongoing basis.

TEAM ROLES

It takes real dedicated work to develop great teams. There can be many challenges. Examples are: if the project manager only has the project participants temporarily; if the project manager holds on to work tasks instead of delegating; or if the project manager lack the knowledge and understanding of what it takes to develop a team.

One tool for ensuring harmonious collaboration is to think about the roles played by the different project participants. Here, it may be a good idea to be inspired by team role theories, which describe the importance of having project participants who fulfil different team roles in the project and their professional roles to create the 'perfect team'.

Working with team roles entails fostering a shared language and mutual respect among members. This allows them to comprehend each other's role contributions and personal preferences, thus helping to prevent major conflicts within the group.

> No one is perfect – but a team can be! Meredith Belbin

Meredith Belbin is widely recognised as the pioneer of the team roles concept (Belbin, 2010). Belbin identified a number of different team roles and mapped the dynamics between the roles and their influence on the overall dynamics of the group. According to Belbin, a team role is a role that an individual assumes in a group constellation that goes beyond our professional role. Belbin defines team roles as one of nine clusters of behavioural attributes identified by his research at Henley as being effective in order to facilitate team progress. Team roles emerge as a mix of an individual's experiences, values, motivation, personality, and learning capacity. Typical team competencies are process oriented. The nine team roles of Belbin are shown in Table 5.

Developing your team involves fostering a dialogue about everyone's strengths and weaknesses to ensure everyone contributes effectively. Engaging in discussions about each person's preferred role can lead to an agreed-upon division of responsibilities, optimising teamwork. Generally, people have two–three primary preferences for specific team roles and tend to excel in the competencies that come most naturally to them.

Too much or too little of the same team role creates imbalances in a team which creates a need for too much energy to rebalance the group. The point is that too many similar team members give rise to too many 'allowable

Table 5 Description of Belbin's nine team roles.

Team Roles	Description	Strength	Allowable Weaknesses	Don't be Surprise to Find That
Resource investigator	Uses their inquisitive nature to find ideas to bring back to the team.	Outgoing, enthusiastic. Explores opportunities and develops contacts.	Might be over-optimistic and can lose interest once the initial enthusiasm has passed.	They might forget to follow-up on a lead.
Teamworker	Helps the team to gel, using their versatility to identify the work required and complete it on behalf of the team.	Cooperative, perceptive, and diplomatic. Listens and averts friction.	Can be indecisive in crunch situations and tends to avoid confrontation.	They might be hesitant to make unpopular decisions.
Coordinator	Needed to focus on the team's objectives, draw out team members, and delegate work appropriately.	Mature, confident, identifies talent. Clarifies goals.	Can be seen as manipulative and might offload their own share of the work. They might over-delegate, leaving themselves little work to do.	They might over-delegate, leaving themselves little work to do.
Monitor evaluator	Provides a logical eye, making impartial judgements where required and weighs up the team's options in a dispassionate way.	Provides a logical eye, making impartial judgements where required and weighs up the team's options in a dispassionate way.	Sometimes lacks the drive and ability to inspire others and can be overly critical.	They could be slow to come to decisions.
Plant	Tends to be highly creative and good at solving problems in unconventional ways.	Creative, imaginative, free-thinking, generates ideas, and solves difficult problems.	Creative, imaginative, free-thinking, generates ideas, and solves difficult problems.	They could be absent-minded or forgetful.
Specialist	Brings in-depth knowledge of a key area to the team.	Single-minded, self-starting, and dedicated. They provide specialist knowledge and skills.	Tends to contribute on a narrow front and can dwell on the technicalities.	They overload you with information.

(Continued)

Table 5 Description of Belbin's nine team roles (*Continued*).

Team Roles	Description	Strength	Allowable Weaknesses	Don't be Surprise to Find That
Shaper	Provides the necessary drive to ensure that the team keeps moving and does not lose focus or momentum.	Challenging, dynamic, thrives on pressure. Has the drive and courage to overcome obstacles.	Can be prone to provocation and may sometimes offend people's feelings.	They could risk becoming aggressive and bad humoured in their attempts to get things done.
Implementer	Needed to plan a workable strategy and carry it out as efficiently as possible.	Practical, reliable, efficient. Turns ideas into actions and organises work that needs to be done.	Can be a bit inflexible and slow to respond to new possibilities.	They might be slow to relinquish their plans in favour of positive changes.
Completer finisher	Most effectively used at the end of tasks to polish and scrutinise the work for errors, subjecting it to the highest standards of quality control.	Painstaking, conscientious, anxious. Searches out errors. Polishes and perfects.	Can be inclined to worry unduly, and reluctant to delegate.	They could be accused of taking their perfectionism to extremes.

Source: with permission from www.belbin.com.

weaknesses', and the group as a whole takes on the weaknesses. Similarly, too many similar team members' strengths can give rise to 'too much strength'. Whether the group has too much strength or too many weaknesses, it too easily leads to conflicts within the group. The team purpose helps to define the needs of the team and the individuals required to fulfil these needs. Also, to form high-performance teams, it's essential to prioritise diversity and balance in project teams. A well-balanced team appreciates the unique roles that each member plays, ultimately forming a strong and resilient collective unit.

Regardless of the team role tool used to evaluate individual members' contributions and preferences, it is ultimately the project manager's responsibility to ensure effective collaboration. When assembling a team, we first assess the project's requirements and objectives, selecting the appropriate skill sets for the tasks at hand. The team is then formed, often based on factors such as availability, professional competencies, and prior project experience.

Sometimes you succeed in creating a team with a good mix of team roles - other times, you don't quite succeed. Whatever the composition, the most important thing is to be aware of the strengths and weaknesses of the team as a whole in relation to the task at hand. In this way, we can succeed in creating the best possible team based on team role theory. Giving up prematurely or pretending that everything is fine can be detrimental to team development, as it greatly relies on fostering trust, nurturing relationships, and understanding each member's preferences and working methods.

In our experience, team members are rarely replaced because the team as a whole lacks certain team roles. Instead, based on mapping the team's strengths and weaknesses, the group will jointly set-up rules or principles for cooperation that compensate for the team's weaknesses. The most obvious approach is to let individual team members use their secondary preferences, knowing that it requires more energy for the individual to bring themselves into play on process competences that are not primary. An alternative is to bring in resources temporarily from outside.

To succeed with the project, it is, therefore, crucial to develop a team with a common mindset and language around the team's strengths and weaknesses; and that the project manager continues to focus on making the group dynamic work well throughout the project's life.

CREATE HIGH-PERFORMANCE TEAMS

Regardless of team composition, each group undergoes a development process that significantly influences how well members get to know and trust one another. What causes some teams to become cohesive units while others remain merely a collection of individuals working independently? Many have experienced the latter or even a middle ground, with an attitude like, 'It's fine; we're getting our work done'. However, far fewer have had the opportunity to partake in a team that truly evolves and grows together. The development of a successful team is often hindered by the diversity of its members, who bring their individual beliefs, preferences, and personal goals to the table.

There are several different theories that explain how composite groups can manage to evolve through different phases. All teams are different, so this is obviously a simplistic explanation in the face of a reality that is full of nuances. Nevertheless, the models suggest a series of activities that the project manager can undertake to develop the team towards the desired behaviour. Psychologist Bruce Tuckmann created a team development model (Tuckmann, 1965) named after the stages the model takes the team through. The model is called 'forming, storming, norming, and performing' and is illustrated in Figure 22.

Forming: shaping the group

The first phase, known as 'forming', involves group members intuitively seeking similarities and shared needs with their teammates. Most individuals will be polite and 'keep up appearances' during this phase. For some, this stage may evoke feelings of insecurity and prompt numerous questions about the why, what, and how. Others might find this phase more familiar, having experienced it multiple times before. For yet others, everything may seem new and thrilling, stirring up a strong desire to simply dive in.

In this phase, the project manager can initiate activities that remove uncertainty, answer questions, and prevent a smooth transition to the storming phase (see Figure 22).

Storming: the storming phase

The second phase, known as the storming phase, is when civility crumbles and the project atmosphere becomes a little more tense – or downright stormy. Some of the team members will focus on acquiring power, others will seek influence,

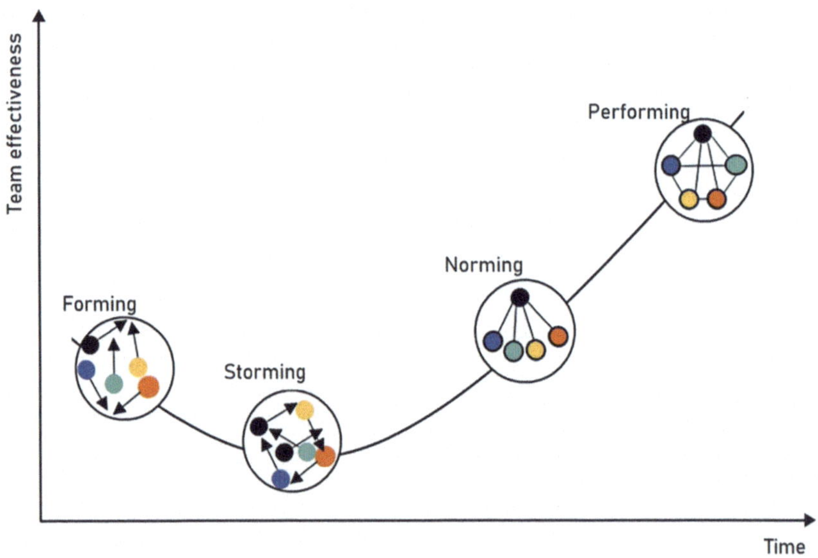

Figure 22 Team development phases over time in the context of team effectiveness (Tuckmann, 1965).

some will challenge the project manager, and some will seek to fulfil their own needs. All in all, a phase where politeness often is gone, and some members are more direct, whereas others become more hesitant or even withdrawn. Unfortunately, this also means that conflicts are lurking beneath the surface. Fortunately, this stage also sees the beginnings of collaborative relationships and team members working towards advancing to the next phase, especially if the project manager is good at navigating everyone through the phase.

Norming: the return to normality

During the norming phase, the group begins to view itself as a unified team, collaboratively tackling various tasks and roles. Members display interest in one another, and the group establishes 'norms' for effective cooperation. These norms typically encompass aspects such as meetings, communication, and behaviour.

Performing: the effective team

Now, the group faces a challenging phase transition following the satisfactory norming stage. Team members must recognise that further growth in teamwork lies ahead. The team will progress into the performing phase, marked by mutual trust and respect among members and a shared sense of accomplishment related to the project's outcomes. It is only at this point that we can discuss sustainable teams – where cooperation is built on trust and care, members support and assist one another. Recognising each other's work is natural, and everyone is prepared to go the extra mile for others to succeed in the task.

Teams vary in dynamics, and their progression through the four phases depends on the project manager's leadership. A proactive and supportive management style from day 1 facilitates quicker achievement of high team effectiveness. The project manager plays a crucial role in maintaining the team's health and balance through tangible activities. It is an advantage if the project manager can perceive the stage of the group energy and has a sense of what is needed to move them forward through the different phases.

There will be stages where some team members may feel insecure, whereas others seem to be 'soaring'. The project manager's responsibility is to guide and support the group's development, ensuring that all team members progress through these phases cohesively and as harmoniously as possible.

Whether a group progresses quickly or slowly through the different phases depends on the initiatives undertaken, the project manager's leadership behaviour, and the individual members' input. Although most project groups can deliver results regardless of their development stage, there can be significant differences in quality, time, and cost, which comprise the project triangle.

When new team members join the group, they will be in the forming phase. It is essential for their integration that all team members briefly revert to the forming phase as well. If new members do not receive a proper welcome, they may never fully integrate, causing the group to lose some of its momentum. By temporarily returning to earlier stages, you can ensure successful onboarding of new members. With a focused approach, the group will quickly regain its footing and return to previous progress levels.

A project manager who neglects team development may struggle to attract members in the long run. To build a sustainable team, a project manager should adopt a long-term perspective that fosters optimal team dynamics. High-performing teams are marked by synergy, empathy, and ongoing self-reflection on individual behaviour and group dynamics. Given the often-limited time available for team building in projects, it is crucial to use planning tools (as discussed in Chapter 2) from the first day. These tools facilitate rapid familiarisation among team members, clarify project goals, establish roles and responsibilities, and strengthen trust within the group, see Table 6 for additional activities to consider.

Trust is a crucial component in team development. The level of trust between individual members can vary greatly, depending on their familiarity with one another prior to the project's inception. Team members who have previously worked together successfully may begin a new project with a high level of trust. Conversely, some individuals may enter a new team with a high level of trust as a default, whereas others may start from a more neutral or even negative position by default. This blend of varying trust levels can create a diverse landscape at the outset of a project.

Typically, trust among team members may decline during the storming phase if the project manager fails to transition the team to the next phase relatively quickly. A temporary drop in trust is hard to avoid in the storming phase, as the project manager should step up to the plate.

Trust can be viewed as an account, with the level of trust reflecting either contributions or withdrawals from that account. If the project manager successfully moves the team through subsequent phases, trust will

Table 6 Description of each phase of the typical behaviour, the project manager's approach to management, and different activities that can be undertaken to support development in each phase to the next phase.

Phase	Typical Behaviour	Leadership Style	Activities
Forming	Pending, searching, polite	Visible, communicating, transparent, involving	Create clear goals for cooperation, use the 5 essential tools, set-up team rules Team building activities
Storming	Power struggles, conflicts, challenging, fulfil own needs	Provide space for discussion, build trust, and common processes for work	Create common routines Conflict resolution Provide goals and direction Support team members
Norming	Group identity, common norms	Delegating, creating space	Coaching team members Create freedom with responsibility and a culture of recognition
Performing	Focus on the task, help each other, and create synergy	Motivating Daily dialogue and follow-up	Celebrating successes Team building activities Create a 'feedback culture'

grow – particularly when appropriate initiatives encourage the group to further enhance cooperation.

A final phase called 'adjourning' is sometimes added to the model. This phase is characterised by the break-up and completion of the project. For some, the end of the project is a kind of a mourning phase. For others it's an upbeat phase. This depends on the individual's approach to the project, personal preferences, and what the individual feels about and associates with endings. During the adjourning phase, it is crucial for the project manager to engage in dialogue with each participant and align expectations for everyone's continued involvement throughout the last time in the project.

BUILDING TRUST

> If you treat people as THEM, as objects, or as 'others', they in turn will treat you instrumentally. David and Maister

Trust, as described by David and Maister, can be represented using a trust equation (David & Maister, 2021). The trust equation consists of three concepts in the numerator and one concept in the denominator, which together in a fraction will express the trust factor or 'level' of your trust account.

The fundamental concept of the equation is that trust plays a crucial role in establishing relationships, and it is rooted in the interplay of various factors (see Figure 23):

Credibility: being credible means that others can trust what you say because of your professionalism, expertise, or skills. That you produce the results you promise and deliver on time. High credibility: 'I can trust what she tells me about …'.

Reliability: means that you do what you say – walk the talk, take responsibility. You keep appointments, and if you can't, you give clear feedback. You are known for delivering good solutions on time. High reliability: 'I can expect that he will …'.

Intimacy: is about the depth of the relationship between you and the other person; being able to talk about important topics – especially, those that have personal feelings attached. It is about listening and creating a space of trust. High intimacy: 'I can confide in her about …'.

Self-orientation: the degree to which you work according to your own agenda or for the agenda of the common good. Where does your attention lie, by what principles are you steering yourself? High self-orientated ring: 'He is most concerned with his own worries and needs'.

Based on the trust equation, there are various methods to cultivate trust in your relationships. To enhance your credibility, demonstrate competence

$$\text{Trust} = \frac{\text{Credibility} + \text{Reliability} + \text{Intimacy}}{\text{Self-orientation}}$$

Figure 23 David and Maister's trust equation.

by working transparently, setting clear expectations, and listening more than speaking. You can foster reliability by display of responsibility, keeping promises, maintaining honesty about your work, and strive to understand the other person's perspective. Building intimacy involves attentive listening, sincerity, focusing on the other's interests and concerns, openness, and sharing personal experiences. Reduce self-orientation – that is, give priority to the results of the team – gives the highest return on investment on the trust factor.

A strong foundation of trust in a relationship leads to greater freedom. Such relationships are marked by active listening, openness, and respect, with both parties going out of their way to protect, defend, and accept each other. However, we all occasionally withdraw from the trust account – sometimes due to unintentional lapses in credibility, reliability, or intimacy, and other times for more deliberate reasons driven by a self-oriented action perceived negatively. This is a natural process; it is crucial to recognise that trust must be rebuilt for optimal results. Trust is a delicate factor that requires ongoing nurturing and care.

THE SCARF MODEL

David Rock, who coined the term 'neuroleadership', is the co-founder and CEO of the Neuroleadership Institute. He introduced the SCARF model (Cox, 2012), which highlights the connection between individual performance and brain function. The core principle of the SCARF model is that by concentrating on effective communication, we can tackle challenges in team collaboration and adapt to changing situations. The model describes that through our communication we should focus on minimising threats and maximising rewards. When we feel threatened or stressed in a situation or change – either physically or socially – the release of the stress hormone cortisol negatively affects our creativity and productivity. According to neuropsychological research, the brain continuously screens for threats in social relationships or the environment. The response is expressed through our emotions and behaviour.

Rock has created the model based on five basic human patterns, with the first letter of the English words naming the model:

- <u>S</u>tatus: if your status is threatened, it negatively affects performance. If you perceive yourself as having a high status in the situation, you are more likely to put yourself at stake.
- <u>C</u>ertainty: if there is uncertainty about your manager's assessment of your performance, this creates uncertainty and negatively affects performance. Conversely, if you have certainty that you are supported and encouraged in what you do, your performance will be greater.
- <u>A</u>utonomy: if you are not allowed to do what you think is right, it will negatively affect your performance. A high level of autonomy will give you room for manoeuvre and natural ability to take independent initiatives.
- <u>R</u>elatedness: if you don't have a good connection with the others on the team, it will negatively affect your performance. Well-founded social contacts naturally create the willingness to contribute to the fulfilment of the task.

- Fairness: if unfair decisions are made, this will negatively affect performance. Some will put more emphasis on fairness and therefore perform better, if the starting point is fair.

Again, it is the amygdala that is at play. When the amygdala is activated, the brain will choose the safest path, activate familiar patterns, and the person avoids new solutions and opportunities. If any of the five factors are negative it will lead to a prompt, rapidly and clear response from the brain, assessing the severity of the threat. Or if the factors are positive, that brain will signal that everything is good. The project manager's insight, understanding, and mastery of the five factors can promote the positive development of the team. If a team is managed based on positive SCARF factors, you will find that there will be less conflict and increased innovation in your team.

Together with your team, you can evaluate the five basic factors and learn more about what each individual needs in different situations in daily life:

Status: How do you perceive your status in the organisation or your status in relation to the others on your project team? What is your understanding of how others perceive you? This can range from insignificant to significant.

Certainty: To what extent can you predict what will happen in the short or long term? It's about clarity and opportunity for the individual and being well-prepared. The project environment can be anything from unpredictable to predictable.

Autonomy: It refers to self-determination and the ability to make one's own decisions. It encompasses having a sense of control over the events that impact you within a project. Being consulted and maintaining freedom in choosing solution methods are essential elements of autonomy, which can range from disempowerment to empowerment.

Relatedness: It relates to one's sense of connection with another person and the team. It is about a sense of security in the relationship(s). Am I in the team, or am I outside? This factor can vary from unsafe to safe.

Fairness: Is there an atmosphere of objectivity and non-judgemental behaviour in the group, especially from the management/project management? Do I feel that what is happening is unfair or reasonable in the light of my own perception of the situation? Again, this factor can vary from unfair to fair.

Utilise the SCARF model to tailor your communication within the team, taking into account each individual's needs and how they relate to the model's patterns. Be mindful of what transpires in the social environment during stressful situations. As you become adept at identifying the five patterns, adjust your communication accordingly. The following is an overview of what to consider and how to maintain equilibrium among these five emotional patterns.

Status: Most people seek to be acknowledged according to their perceived self-worth. Generally, individuals desire equal footing in conversations and mutual recognition of each person's work and contributions to a project. However, those who crave status as a social element prefer emphasis on their accomplishments rather than focusing on mistakes or shortcomings. A good rule of thumb is that most people do their best they can in any situation.

Certainty: The human brain favours a reasonable degree of predictability, so you should always ensure that all participants have the right and sufficient information. We organise the project to establish clearer communication lines, which helps some people. We typically communicate too little in projects. Mostly because we think we've said it or because it's so obvious to ourselves that it doesn't need repeating. Continuous communication throughout a project ensures that participants have the necessary knowledge, allowing them to feel confident and secure in their daily tasks.

Autonomy: Most individuals desire involvement in decision making to exert influence. Acknowledging and valuing participants' input, as well as delegating crucial tasks, can enhance the group's sense of autonomy.

Relatedness: Human beings are inherently social creatures who desire a sense of belonging within a group. Consequently, for some individuals, forming relationships within the group holds great importance, and they often prioritise the human aspects throughout a project.

Fairness: Our reactions are triggered when an event challenges our sense of fairness. Some individuals may contribute more effort to the group without receiving additional compensation, leading them to perceive the situation as unfair. In certain cases, one group of stakeholders may be impacted more significantly by a project than others, resulting in feelings of unfairness. To address this, we must promote transparency along with open and honest communication about the changes emanating from the project.

CONFLICT IS COMING

Sometimes it seems as if conflicts arise out of nowhere, but often it is a sign that one or more of the parties have not recognised the conflict until it has escalated to a high level on the conflict ladder. There are many people involved in project processes and product development, and we have to be almost supernaturally skilful to avoid conflicts - even if we work continuously to create balance. Conflicts are not necessarily purely negative. If resolved properly, they can emphasise the changes brought about by the project, strengthen cohesion, and renew commitment. On the contrary, there is no doubt that prolonged unresolved conflicts steal energy and resources from project staff and contribute to delays, cost overruns, or lack of quality in project deliverables. Most of us remember conflict-ridden projects and how hard it can be to be surrounded by arguments, bad feelings, stress, or a bad atmosphere.

Most conflicts arise within the project team between project participants, stakeholders, and partners. The theme of conflicts is everything from resistance to change in relation to the project's end result to frustrations about time, quality, and economy, as well as choice of the project's technical solutions. Conflicts are more likely to arise where there is a lack of trust. If trust is present, cooperation can often accommodate disagreements, and new ideas and possibilities can emerge. The most dangerous conflicts are those where parties seem to agree superficially but harbour underlying mistrust. These hidden conflicts can escalate unnoticed before they suddenly erupt, often catching one party by surprise.

As a project manager, it's crucial to stay attuned to potential conflicts within your team. Common indicators of conflict include silence, arguments, avoidance behaviour, gossip, backstabbing, and physical symptoms such as tears, stomach aches, heightened vocal pitch, and stress. Most individuals can recognise their own unique signs and symptoms of conflict. As a project manager, promptly assess the scope and severity of any detected conflicts. Assessing the status, extent, and escalation of the conflict is essential. If you do not respond promptly, the conflict can quickly slip out of control without you realising it, and you may inadvertently become a participant in the conflict.

When conflict is detected firstly, avoid drawing conclusions or making interpretations about the conflicting behaviours. Instead, gain an overview of the involved parties, the chronological sequence of events, and the root of the conflict. Is it a misunderstanding, differing interests, or a clash of values? Often, conflicts stem from unclear responsibilities, lack of competencies, or misaligned expectations among parties. Unmanaged risks, overly optimistic schedules, poor communication, or inadequate cooperation can also contribute to conflicts. Additionally, disputes may arise due to contractual content, requirements, understanding, and interpretation.

Easier conflicts to resolve typically involve misunderstandings about solutions, cooperation, and similar situations. On the contrary, more challenging conflicts delve into human values and power dynamics. Whether dealing with light or heavy conflicts, resolution is usually possible to achieve through dialogue between the parties involved.

Before initiating dialogue in conflict resolution, it's helpful to consult the 'conflict ladder', which illustrates psychological and behavioural aspects at various conflict stages. The conflict ladder also suggests the most suitable forms of dialogue for each stage (see Figure 24) (Monberg, 2019).

In the first stage, disagreements arise, and opinions are shared. The focus remains on the issue at hand, with both parties willing to work towards a solution. Luckily, during this phase, numerous potential solutions can be found that cater to the interests of both parties. It is essential that mutual trust is maintained and that the problem is kept separate from the people involved. In practice, many conflicts within projects reach stage 1 and are resolved at this point.

If trust declines or the solution fails to benefit both parties, the conflict may escalate to the next stage. This happens when one or both parties shift their focus from the issue to the person. As trust diminishes, the dialogue can quickly turn into an argument, centred around advocating one solution over another. Stubbornness at this stage is usually unproductive, and returning to stage 1 requires openness, flexibility, understanding of each other, and a mutually beneficial solution. Prolonged time in stage 2 leads to blame-shifting and heightened emotional involvement. Consequently, mistrust grows and pushes the conflict into stage 3.

Here, the dialogue will be characterised by the parties perceiving the other party as dishonest and action to resolve the conflict will be absent. However, there may be a small chance of resolving the conflict at stage 3, which requires the involvement of a third party. The third party can help the parties down the stairs by using neutral and open questions to clarify the conflict.

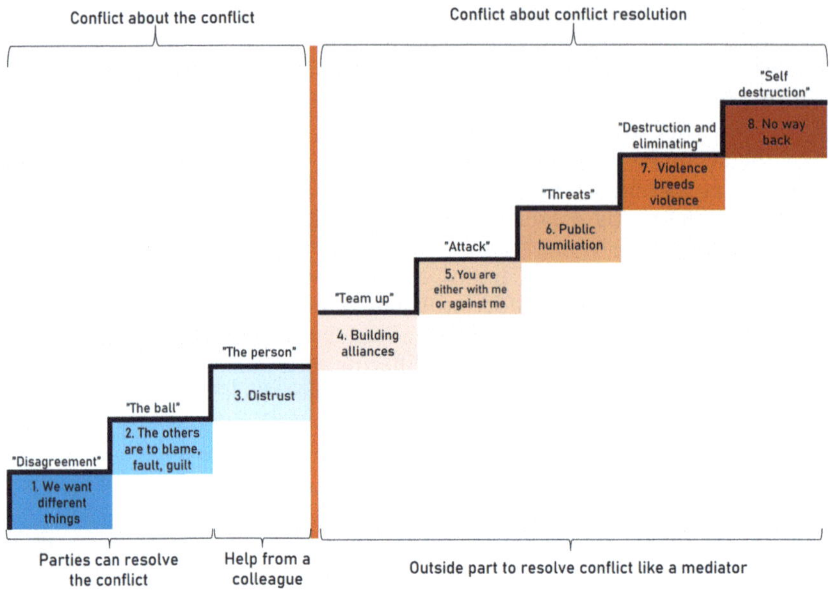

Figure 24 Conflict staircase showing typical signs of each stage of escalation (Glass as in Monberg (2019)).

If the conflict escalates, it's usually because the dialogue halts at stage 3. In stage 4, the involved parties begin seeking support from others and forming alliances to validate their stance. However, such alliance-building can prove detrimental for everyone, as colleagues are forced to choose sides under the 'either you're with me or against me' mentality. Consequently, the conflict expands beyond its initial scope and drains the organisation's energy. At this stage, resolving the dispute requires considerable effort from a project manager. Typically, it necessitates an unbiased external party who works collaboratively with those in conflict to find a resolution.

From stages 5 to 8, the conflict escalates, becoming increasingly severe, with notions of morality and ethics potentially fading. The involved parties may fear losing face and focus on obtaining an apology from the opponent and/or demeaning their opponent. Both sides might resort to threats, humiliation, and even violence in the final stages. As indicated earlier, conflicts at this stage are typically challenging to resolve without external assistance by a mediator. If a project manager identifies such a conflict, they should seek help from a mediator who can employ the mediation process to address and resolve the conflict.

The conflict staircase illustrates the escalation of conflicts and which methods of resolution are effective at the different stages.

Communication is crucial when dealing with situations that can develop into conflict and in conflict situations. When communicating, we can focus on the words we use that either escalate or de-escalate conflict (see Table 7).

Table 7 Escalating and de-escalating language.

Escalating Language	De-Escalating Language
Going after the person	Focused on the case/action
Focuses on errors and omissions	Focusing on hopes and dreams
Focusing on the past	Focuses on present/future
Will be right	Seek mutual understanding
Placing blame	Taking responsibility
Speaking in 'you' form	Speaking in 'I' form
Attacking, judging, and assessing	Exploring and listening
Coercive	Engaging
Making demands or defending self	Making room for diversity
Generalising (always, never)	Requesting or wishing
Interrupting	Listening

Escalation occurs when we attempt to convince others that our perspective is correct, focusing primarily on our own opinions and arguments. Another form of escalation is using a kind of language in which we tell others what they are feeling and thinking. Consequently, the other party feels attacked and becomes defensive, decreasing the chance for open dialogue. Moreover, focusing on 'what happened' and the causes only serves to heighten the conflict.

De-escalating language, conversely, relies on a more 'circular understanding' to gain deeper insight into issues. This approach emphasises the exploration of connections and patterns, encompassing thoughts, relationships, and actions, as well as the outcome and perception of those actions. Conflict resolution necessitates continuous dialogue and mutual comprehension of each party's needs and interests.

Mediation serves as a structured method for resolving conflicts at any level. The mediation process is detailed in Chapter 9.

ASKING THE GOOD QUESTIONS

One effective method for relaxation and conflict resolution is utilising skilful questioning techniques. Well-crafted questions can encourage dialogue, stimulate reflection, and promote critical thinking, making them valuable in situations that require problem solving, conflict resolution, and team coaching. Optimal questions are 'open-ended', fostering a built-in curiosity and understanding through dialogue.

Asking effective open-ended questions is a learnable skill. Traditional questioning techniques often rely on linear thinking, progressing from problem to solution. However, utilising a circular questioning approach can yield better results, both in terms of solutions and fostering a positive atmosphere among participants. With this method, the focus remains on the relationships and various hypotheses present within the situation. Using linear questions narrow down the space of solutions, whereas using circular questioning opens up the situation for more hypotheses and solutions.

An experience was shared by a project manager who once confided about having a strained relationship with a specific subcontractor, which led to unfavourable outcomes. One day, he chose to adopt a new approach and questioning technique during their discussion about the subcontractor's work. Within an hour of their meeting, the dynamic between them shifted to a positive, friendly conversation based on the modified questioning strategy alone, and a shared goal of fostering productive collaboration moving forward.

In the 1980s, Penn and Tomm developed a model for effective questioning (Tomm, 2022). This methodology is grounded in systemic and appreciative approaches. Furthermore, the model employs circular learning assumptions, promoting increased reflection and deliberation before settling on solutions.

The questioning technique asks questions from four different starting points (see Figure 25):

(1) The past-oriented simple questions to clarify the situation and create mutual clarification of situations (what, who, and when).
(2) The past-oriented complex questions will help you to put the situation into perspective and expand and develop to gain understanding of the situation (experiences, influences, what do others think – a circular approach to the subject).
(3) The future-oriented complex questions focusing at creating possibilities and new ideas (advice, hypotheses) and to create reflection.
(4) The future-oriented simple question has an influencing perspective and clarifies actions and look at options for action (first step, what do you do now, who can help?).

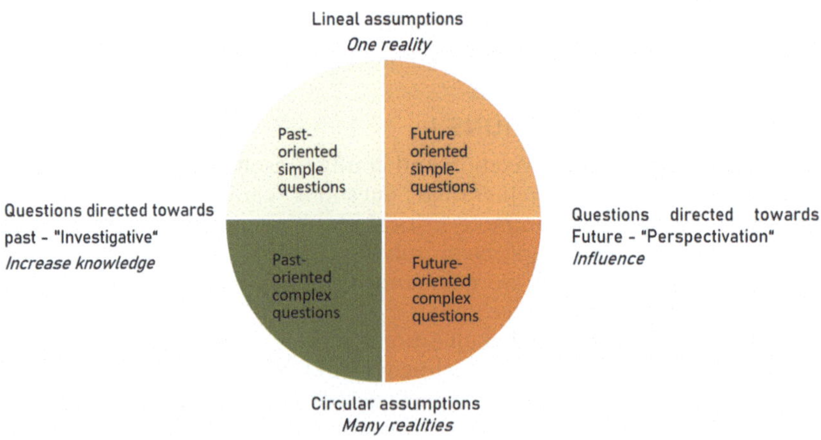

Figure 25 Circular questioning technique (Tomm, 2022).

Phase 1: Past-oriented simple questions

The primary goal of the first phase is to explore and define the background of the issue while clarifying relevant facts. Exercise caution in this phase to avoid asking excessive questions, as this can lead to repetition or closed-ended questions that yield no new insights. The questions should focus on past events, and this phase's emphasis should be on clarifying misconceptions, resolving misunderstandings, and gathering sufficient information about the situation for a comprehensive understanding.

Good linear questions in this phase are, for example:

- What is the problem?
- How long has it been like this?
- What has happened?
- Who else is involved in the situation?
- How do you usually deal with similar situations?

Phase 2: Past-oriented complex questions

The aim of this phase is to examine the relationships, positions, and interconnections. Questions in this phase will primarily address past events but will focus on context and patterns associated with the situation. These questions will attempt to gather different perspectives and explore whether new connections arise and have a circular approach to the subject. Another crucial objective is to challenge the mental assumptions inherent in the situation.

Good circular questions in this phase are, for example:

- How do you think X sees the situation?
- What is the difference between the way you and X see it?
- Who else sees the situation differently than you?
- When is/was the challenge greatest/least?
- What makes you think that?
- When and in what way will the challenge be easiest to solve?

Phase 3: Future-oriented complex questions

The intention of this phase is to challenge the focal person to find new perspectives and see new contexts, challenges, mental models, and so on. The questions in this phase will be directed towards the future and the answers will be different scenarios and hypotheses. This phase aims to generate fresh ideas, discover alternative possibilities, and envision new pathways. Some suitable flexible questions for this stage might include:

- How would X think and/or act if X had the same challenge?
- What would you have done if this had happened 2 years ago?
- What would happen if the challenge took centre stage?
- What will happen if you do nothing?
- What will it be like when the problem is solved?
- How do you know that the problem has been solved?
- Can you imagine that …?
- What can you do differently from now on?

Phase 4: Future-oriented simple questions

In this phase, the aim is to promote goal setting for altering thought patterns and/or behaviours, and to extract valuable lessons from the conversation. This stage focuses on defining new initiatives and inspiring action. The questions posed will elicit responses that clarify and conclude. Effective action-clarifying questions can be:

- What do you need to do to put your new knowledge into action?
- Who or what can support you in your endeavours?
- What is the first step?
- What new initiatives will you take?
- What can help you/support you in this?
- What can stop you from acting? Why?

Remember that the aim is to get the best out of the situation and ensure that the right solutions are brought to the table. The very best questions are the ones we ask where the other party needs to think before answering.

Closed questions offer limited insights, as they only require a yes or no answer. On the contrary, open questions foster new ideas and viewpoints. Employing open questioning techniques can aid in discovering innovative solutions for projects, facilitating team members' growth, and resolving conflicts.

The circular questions model can be applied in various situations, with the most significant being to encourage the exploration of complex questions and gather the essential knowledge as a foundation for discussing different perspectives and finding solutions in the project and within the project team.

MAIN POINTS

In team settings, we typically adopt two or three primary process-oriented skills, which can be identified as team roles – distinct from our classical professional expertise. To form an ideal team, the project manager should initiate discussions with members about their preferred team roles. It is crucial for the project manager to balance the team by appreciating each member's unique role contributions and establishing common ground rules to facilitate smooth collaboration despite individual differences.

Team development progresses through various phases, and it is the project manager's responsibility to implement initiatives throughout these stages to ensure optimal support for the team. This enables them to achieve the best possible results. A sustainable project manager strives to create cohesive teams characterised by synergy, mutual respect, empathy, and self-reflection. Such teams possess the ability to accomplish extraordinary feats on their path to success.

Trust is a crucial component in all cooperative relationships, including team development. Although building trust requires effort, it can also be lost quickly. To lead their team to success, project managers must earn trust and initiate trust-building activities. They should demonstrate credibility and reliability while fostering a confidential environment without being overly self-centred. By doing so, they can achieve a high level of trust within their team.

Focusing on the social environment and communication within the team can enable project managers to prevent resistance and conflicts. By being mindful of five fundamental factors – trust, safety, autonomy, relatedness, and fairness – you can foster a positive mindset within the social environment of the team and make major contributions to the team's healthy development.

Throughout the duration of a project, most project managers frequently encounter conflict. Trust-based team management can create an environment in which conflicts are effectively managed. Nevertheless, when disputes arise, it is crucial for the project manager to intervene and ensure that they are resolved in a healthy and balanced manner by employing de-escalating language and dialogue.

YOUR REFLECTIONS

Project teams often come from different organisations and social environments. Try to conduct a status analysis on a project team you are working with now.

Focus on team roles

Can you identify the preferred team roles of each project participant in your team? Does the team cover all team roles so that you have a balanced team? Can you identify the preferred team roles of each project participant in your team? How can you strengthen the team where certain team roles are missing?

The team development process

Reflecting on your past projects and project teams, was there a particular team that excelled as a 'performing team'? What factors contributed to this outcome? What actions did you personally take that were successful? Is there any of the models from this chapter that can be used to explain some of the success of the team?

Do you have a well-functioning team now? Can you determine which phase the team is in? Have you already initiated activities relevant to the phase? Are there any other activities you can initiate that will develop the team further?

In what way do you work with trust?

How do you assess the level of trust to be in the team as a whole? How big is the trust factor between you and each of the project participants? Can you sense whether the team trusts you in general? How high is the trust factor? What can you do to change the trust between you and the team? Are there relationships in the team where there is a deficit in the trust account? How can you help change the trust factor in those relationships?

The social environment of the project

Conduct a writing exercise in which you reflect on the five factors of the SCARF model: status, certainty, autonomy, relatedness, and fairness. Consider how you imagine that the different team members perceive the state of the project on each factor. Is there anything that could be improved? Do you get annoyed by someone in your team reacting to one of the five factors? Does it differ from your own approach to the social environment? What can you do to create a better social environment for the participants in your team?

Think about your project and the culture of conflict in the team
How do you feel about conflicts? Are you aware of them approaching? Do you recognise them but avoid addressing them? Regardless of your approach, it's essential to examine conflicts within your team – big or small. Can you identify a brewing conflict? Who is involved? What is the reason? Think about where the conflict is on the conflict ladder. How will you resolve the conflict? What conflict language characterises your team – is it escalating or de-escalating?

READ MORE

'Team Roles at Work' by R. Meredith Belbin, which describes the different roles and how they contribute to the team (Belbin, 2010).

'Power in Projects, Programs and Portfolios' by John Ryding Olsson, Mette Lindegaard Attrup, and Niels Ahrengot (Olsson *et al.*, 2019), which describes the basic elements, tools and experiences in project management, projects and portfolio management.

'TEAM – The Neuroscience of Psychological Safety' at www.neuroleadership.com (Cox, 2012), which deals with various issues such as teams and leadership, and how we collaborate with trust and build psychological safety across organisations and teams.

Karl Tomm and the four types of questions: https://www.youtube.com/watch?v=AGiTJ-Z_y6Q&t=2s (Tomm, 2022) can be used in many different situations. As a project manager, you are in situations as a coach, facilitator, or conflict resolver, where you can use the questions to achieve better results than if you have a linear mindset.

Tina Monberg has worked with conflict management and mediation and has written numbers of e-books about conflict management, about seeing conflicts as positive in an organisation that is willing to change and share and having a conflict strategy with clear guidelines for resolving conflicts. Instead of being afraid of conflicts, the organisation must have clear guidelines for conflicts and ensure that conflicts do not develop negatively. In the book 'Handbook of Human Conflict Technology', you will learn more about how to resolve conflicts through mediation, a tool that is about getting the conflicting parties to see the benefits of cooperation rather than blame and is based on the win–win philosophy (Monberg, 2019).

'Whole Brain Living. The Anatomy of Choice and the Four Characters that Drive Our Life' (Taylor, 2022) is a book about understanding the connection between our thoughts and our emotions so that we no longer have to be bound by our emotional reactions. The book is about being aware of the connection and choosing to live to be the best version of ourselves and control what's going on in our heads.

The book 'Non-Violent Communication: a language of life' by Marshall B. Rosenberg (2015) introduces the non-violent communication also called giraffe language. The book provides an introduction to and experience of using the language.

doi: 10.2166/9781789063349_0197

Chapter 8

Partnerships

Alone we can do very little, together we can do a lot.

Helen Keller

A study conducted by McKinsey reveals that 98% of construction projects exceed their budget by more than 30%, and 77% of them face delays (Changali *et al.*, 2015). Regardless of their complexity, typical challenges in construction projects often result in budget overruns, schedule delays, and unfulfilled deliverables in terms of quality. These findings are consistent across various studies conducted over the past decade (see e.g., Leeds, 2017). Based on these lists and our experience, we have compiled the most prevalent challenges faced by the construction industry, which can be summarised as follows:

- Lack of structure and project objectives
- Inadequate risk management
- No alignment of expectation with owner and other key stakeholders
- Participants with lack of basic project performances
- Cash-flow issues
- Low productivity and profitability
- Shortage of skilled labour
- Poor communication
- Sustainable concerns

Unsurprisingly, most of the challenges mentioned have already been addressed in Part 1 of this book. This reinforces the importance of embracing the available project management tools and continually developing our approach to project management work and increase the project performance.

Regarding low productivity and profitability, it appears that excessive time is spent on waiting for materials and equipment, travel time, extended breaks, and work replanning. A study by the Construction Owners Association of America found that up to 63% of labour time is wasted.

Authors: Pernille Ingildsen and Mette Thorkilsen

In 1979, Glenn Ballard and Greg Howell established the Lean Construction Institute (LCI) to enhance construction site performance based on production system design and management principles, as they identified similarities between the challenges in the construction and manufacturing industries (LCI, 2023). They developed a concept for managing construction projects inspired by lean principles from the Toyota Production System (Liker, 2021). The lean construction system emphasises collaboration and learning and has introduced a planning method known as 'the last planner system', which is a concept on how to monitor and follow-up on the execution of the construction work to minimise waste.

Sarhan describes that the UK construction industry has had difficulties to implement the lean construction concept due to structural and cultural barriers, as well as a lack of top management awareness (Sarhan, 2013). We believe that construction industries in European countries likely share similar cultural aspects with the UK industry. Traditionally, the construction sector has been characterised as an 'execution culture', focusing on hands-on work and going the extra mile. Consequently, systematic planning has often been neglected. Planning phases are frequently not conducted thoroughly due to time constraints. However, most project participants wish to allocate more time for better planning, and the LCI provides methods and systems to achieve this.

In many instances, poor project performance in the construction industry can be attributed to a combination of factors. To address some of these issues, partnering or forming partnerships that involve early engagement from all parties may prove beneficial. By focusing on practical solutions, this collaborative approach can enhance the planning process and streamline the execution phase. Furthermore, partnerships can foster a unified approach for project execution, emphasising shared objectives and promoting cooperation throughout the entire process.

Partnering has successfully been applied in the construction industry for at least a decade. Originated in the USA, it has since been widely adopted in the UK and, increasingly, in other parts of the world, including mainland Europe. When partnering was introduced in the 1990s, it emerged due to contracts failing to fulfil project goals (Scott, 2000). The concept of partnering emerged from the realisation that traditional contracting methods for construction projects often failed to deliver acceptable results for both owners and contractors. Several reasons for this failure have been proposed, each with varying degrees of validity. Fortunately, it seems that the partnering form of cooperation has regained focus over the past 20 years. Nowadays, long-term 'strategic partnerships' are becoming increasingly popular within the construction industry.

In addition to numerous challenges, the construction industry must now also consider sustainability, as new environmental regulations and expectations arise. The industry generates significant carbon dioxide emissions through operations and material usage, making it a major global pollutant. Consequently, there is a need for renewable, recyclable materials that simultaneously enhance energy efficiency. This calls for significant changes in project management, design development, and construction processes for the industry's future. As companies adapt to varying requirements and regulations enforced by governments worldwide, they will face a steep learning curve. Addressing existing challenges related to management, cooperation, and communication is essential, which necessitates innovative ways of interacting within projects.

Facing the future with innovative technologies, new construction methods, ecofriendly materials, and an overall focus on sustainability, the construction industry must adapt its practices. We believe that to create meaningful changes in projects, we need to foster trust, collaboration, and a more unified approach. This chapter explores how to establish partnerships and cultivate a win–win mindset, which we consider to be crucial for future cooperation models. Additionally, the chapter concludes with practical tips for implementing partnerships as a collaborative strategy in projects.

PARTNERING

Partnering, a form of collaboration, originated in the 1960s within the US military. This approach aimed to forge strategic cooperation agreements with key suppliers for complex, high-risk projects necessitating close technological collaboration. Centred around trust-based collaboration, partnering seeks the most effective and economically advantageous solutions. The emergence of project management tools in the 1950s in the USA led to a significant overlap between partnering and project management primarily.

Partnering is a project-based collaboration built on trust and established shared objectives. The term 'partnering' signifies close cooperation, emphasising dialogue, trust, and openness, with early involvement of all parties. To achieve success, the common goal must be clearly defined by all engaged parties through joint activities and a foundation of mutual economic interests.

Basically, the idea is that partnering adds an extra valuable dimension to the classic customer–supplier relationship, where customers order and pay, and the supplier finalises and delivers the product. Partnering is designed to provide the customer more influence on the product, and the supplier will be able to provide greater customer satisfaction, as products are defined in collaboration, and costs are reconciled on an ongoing basis. Partnering is a form of cooperation based on dialogue, trust, and openness.

A central part of partnering is to optimise the project's deliverables through early involvement of the contractor parties, thus bringing more practical hands-on competencies into play in the project development and ensuring that the solutions are buildable and optimally aligned with customer needs and importantly that the economy is balanced and proactively managed throughout the project.

At the start of the project, a set of common objectives are defined and developed together among the actors, that is, what is most important, what defines our success, and what are we to deliver together. The common objectives must be unambiguous and understandable for all parties. Objectives can be supplemented by values or more practical rules for cooperation that express the parties' desire to cooperate.

Partnerships often ensure that economic interests are aligned between the parties. Focus is on the fair and wise distribution of risks and rewards in the cooperation. Rewards can be incentives to create better and cheaper solutions or bonuses that are tied to progress, quality, resources, or other factors. In cooperation, work is often carried out with transparency about the finances of the project, which can be based on 'open books' where all costs are presented to the parties. The intention behind this is to create financial transparency to ensure trust and the best value for money.

The starting point for a partnering collaboration is that cooperation across disciplines and organisations helps to create a better flow in the project and on the construction site, more efficient execution of the project, and better well-being for all parties. To make partnering a success, joint activities in the form of kick-off workshops and social activities are organised so that all participants have a feeling of belonging as part of the collaboration. For collaborative endeavours, participants can establish shared procedures for preventing and managing conflicts, as well as for facilitating effective communication.

Successful partnerships demand a focus on cultivating the appropriate mindset, collaboratively developing solutions, maintaining transparency about finances, and working together to build a healthy social environment and strong personal connections.

Lessons learnt from the construction of the Øresund Bridge

> As a project owner we are getting the co-operation partners we deserve.
> Peter Lundhus, former technical director at the Øresund Fixed Link Project (between Denmark and Sweden)

The motorway and railway connection between Denmark and Sweden, opened on 1 July 2000, was one of the largest Danish construction projects in decades. The Øresund Link represented the biggest infrastructural investment of its kind in Europe at that time and, as part of the Trans European Network Programme, it received European Union subsidies. The project faced strong political opposition and extensive public interest fuelled by environmental groups such as Greenpeace. The immense interest was attributable to its location in a densely populated area and the potential impacts on the sensitive local and regional marine environment. Consequently, the owner – The Øresund Consortium – had to find ways to address these diverse interests to prevent halting or reversing the project during its construction phase.

The owner prioritised three main aspects: fostering a constructive partnership with all parties involved in their management strategy, incorporating environmental concerns into the design and construction methods, and establishing open and transparent communication with stakeholders through advanced hydro-informatics to facilitate dialogue.

The owner applied partnering principles, leading to the successful completion of the project with a well-maintained project triangle. It was even finished ahead of schedule. One strategy employed involved joint planning of execution work between the client and all contractors. This resulted in an optimised execution schedule, enabling all parties to deliver their work 6 months early. An Arthurian 'one for all and all for one' approach was used in negotiating the schedule, creating a win–win situation that benefited everyone through an earlier handover. The client's most significant advantage was the earlier-than-anticipated commissioning of a fixed link between Copenhagen and Malmö, generating additional revenue.

Consequently, strong emphasis was placed on quality and transparency, along with broader dissemination of environmental information, particularly in a readily understandable form. One of the main concerns was the hydrodynamics of Øresund and the impact of the fixed link, making hydro-informatics crucial. The environmental monitoring and modelling programme was designed as a feedback system to quickly return data to the construction process via the hydro-informatics system (Lyngby, 1999). This system comprised field work data and modelling results, offering significant advantages to contractors and others involved in construction. By using hydro-informatics data alongside global positioning systems, construction works were able to proceed efficiently, minimising wasted time when placing rock armour, tunnel elements, bridge piers, and other structural components (Thorkilsen, 2001).

The owner's strategic approach led to the successful completion of the Øresund Fixed Link project. It was built to specification, causing no harm to the marine environment and fulfilling all environmental conditions. Moreover, it was constructed within its original budget and completed 6 months ahead of schedule. This accomplishment reflects not only on the involved parties but also on the effective partnership cooperation and the combined efforts of a comprehensive environmental monitoring programme alongside hydro-informatics as a key discipline.

In summary, the owner of the Øresund Bridge project adopted a comprehensive approach from the start, addressing various issues crucial for the project's successful completion.

- There was a well-defined strategy from the owner's side outlining how, when, and in what manner to establish collaboration among the owner, consultants, contractors, and authorities.
- A procurement strategy focused on the 'most economically advantageous' approach, with clear expectations for contractual cooperation.
- The owner took a proactive approach, aiming to always be one step ahead in construction and the construction process while clearly communicating interfaces, risk-sharing, project plans, and joint planning.
- Stakeholders, including politicians, interest groups, and the public, were taken seriously with all relevant data made available to the environmental authorities daily.
- A trust-based quality strategy was implemented through joint audits and supervision of the construction site for both contractors and clients. This allowed them to have a shared understanding of situations and challenges as they arose.
- Contractors were encouraged to take a proactive approach in addressing problems and unexpected events.
- The owner organisation's roles and responsibilities were clearly defined with established communication channels. Regular joint activities were organised to foster a positive social environment.

- Resources were allocated for communication based on a strategy that emphasised clarity, timeliness, and transparency with project stakeholders. All the client's staff members were carefully selected to ensure that the right competencies were brought into play.

In the Øresund project, the essential project management tools and partnering principles were systematically applied and, in our judgement, were a key reason for the project's success.

Lessons learnt from partnering in 2000s

Throughout the 2000s, numerous collaborative projects took place in Denmark. One of the most significant projects was the construction of a new headquarters for the Danish National Broadcasting Company (DR BYEN). The experience gained from this project revealed that participants faced difficulties in adapting to the cooperation concept and understanding the necessity of establishing clear agreements with well-defined responsibilities. Despite these challenges, there were several positive takeaways from the project, which also demonstrated success in fulfilling the project triangle to some extent, as evaluations exist for us to grow and continuously learn from them.

For a successful partnership, the client must be well-prepared and have a clear vision for the collaboration.

Positive experiences gleaned from the construction of DR BYEN include:

- The owner had identified its own needs and established user processes.
- In the project, there was a well-defined project triangle, well-formulated description, and communication.
- Information on quality level, temporal scope, and budgetary constraints.
- A well-defined tendering and contracting strategy – a form of contracting that fits the profile of the project.
- The partnering agreement included a model for incentives and bonuses.
- Common descriptions of project organisation and roles/responsibilities.
- A joint project office was established for all partners and several social activities were organised.
- The developer provided partnering and lean construction facilitators to promote building processes.
- The client established common site facilities and safety coordination.
- The client created a good working environment by implementing a safety concept that should be followed by all actors on the construction site.

As demonstrated in the Øresund Fixed Link case, it's essential to plan the partnering set-up before signing contracts with parties. Additionally, a high degree of involvement from all parties is crucial for successful partnering. Although it requires hard work, in our opinion, it's a worthwhile investment that leads to significantly improved results in construction projects.

Winds of change

Over the last 5–15 years, an increasing number of industry players have invested time and resources in partnering and new way of cooperating. As a result, they can now engage more proactively in various forms of collaboration. We believe this skill set will soon become essential for successfully delivering positive outcomes in sustainability projects for the future. Adopting this approach is favoured by the emerging generation who find it difficult to flourish in the old outdated confrontational cultures. It's time to act; those who do not prioritise the development in partnering capabilities risk falling behind their competitors in the coming years.

Changes of this magnitude require time, and adjusting to new ways of working can be challenging. Furthermore, maintaining focus on such a significant transformation is essential for long-lasting impact. One issue with introducing new collaboration models in the industry is that, if focus is not maintained, some participants tend to revert to their traditional methods of collaboration and governance. Naturally, resistance to change may arise against new initiatives, and for some, the resistance will be rooted in the fact that the investment is too great in relation to the earnings. The parties ask natural questions such as: What if we lose money? What if we don't have the competencies? How will we succeed in changing our culture? The way to ensure that the new forms of cooperation take root is for all parties to proactively take on the challenge of preparing their own organisation to be able to enter partnering and other partnership-like forms.

STRATEGIC PARTNERSHIP AGREEMENTS

The current trend involves companies seeking long-term relationships and establishing frameworks for such cooperation, which in turn leads to new organisational forms. We are witnessing an increase in strategic partnerships, where contracts spanning 4–6 years are signed to execute a series of projects. Whenever possible, the same project organisation is employed throughout the project portfolio.

Strategic partnerships, unlike traditional partnering, adopt a longer-term perspective. Due to the repetitive nature of the projects, it becomes possible to learn from each endeavour and continuously optimise collaboration. Cheating is de-incentivised, as it destroys the teamwork model – no one wants to partner with an untrustworthy party. This long-term outlook fosters trust and establishes an efficient framework for planning, designing, and executing a project portfolio.

Partnership agreements are best incorporated in a holistic organisation where each party contributes to the collective. Furthermore, resources must be allocated for ongoing innovation and product development to benefit individual projects in the portfolio. Partnership agreements typically include the following elements:

- Portfolio management for projects, ensuring pipeline optimisation for execution.
- A steering group composed of management from all parties, overseeing partnership management.
- Day-to-day partnership management, including project portfolio administration.

- An innovation group dedicated to identifying potential solutions in projects through continuous learning.
- Clearly defined roles and responsibilities for project managers, construction managers, and others.
- A process leader skilled in partnership development, facilitation, and change management.
- A unified organisation with collaboratively set-up.
- A transparent financial model with profit and loss elements and profit-sharing provisions.
- Standardised processes for stakeholder identification, risk management, and communication with external parties.
- Uniform procedures for scrutiny, solution selection, decision making, and more.
- Continuous evaluation of partnership components such as onboarding processes.
- Established protocols for resolving conflicts and disagreements.
- Structured procedures for operational involvement during project handovers to operations members.
- Joint staff training initiatives, team building exercises, shared project offices, and more.

WIN–WIN

Win–win is a philosophy that centres around the idea that two parties can both achieve success if they collaborate – 'it's not you or me; it's both of us'. Additionally, it is grounded in the belief that there is enough for everyone.

Mary Parker Follett was a management consultant and pioneer of organisational theory and behaviour who valued people as far more important than industrial aspects. Although she did not invent the concept, she gained recognition in the 1920s for describing win–win outcomes in conflict situations (Peek, 2023).

William Ury and Roger Fisher worked with the concept of win–win in the 1980s in their description of negotiations (Ury, 2011), focusing on negotiations that yield mutually beneficial outcomes for all parties involved, resulting in satisfaction and win–win situations.

Win–win is also known from game theory, where win–win is defined as a scenario where a mutually favourable outcome is achieved. The theory states that if win–win is not realised, it is a lose–lose scenario by default, as all parties in cooperation lose if win–win principles fail.

American mathematician John Forbes Nash was awarded the Nobel Prize in 1994 for his contributions to game theory (Nash, 1950). Nash's theory posits that each player's strategy is optimal when considering both their own decisions and those of others. All players can achieve their desired outcomes, providing trust is maintained within the group. Should one party change their strategy to gain an advantage, it would disrupt the collaborative nature of the game, leading to loss for others. The term 'Nash equilibrium' refers to a situation where individuals have no incentive to change their strategy, as they perceive their current approach as optimal considering their opponent's choices. In such circumstances, changing one's strategy while others adhere to theirs results in no clear winner.

A risk in partnership collaboration is that initially, everyone involved in the project is enthusiastic and agrees on how to finalise it. However, as the execution phase nears completion, some participants may change their approach due to time pressure or financial constraints. This can lead to a drawn-out process of rectifying defects that leaves all parties dissatisfied, ultimately resulting in failed partnership and unhappy users.

The Prisoner's Dilemma is an example of a Nash equilibrium. The Prisoner's Dilemma was developed by Merrill Flood and Melvin Dresher in 1950.

Two people commit a criminal offense together and are arrested by the police, who have no evidence of the criminal offense. The two prisoners are separated to prevent communication between them. They are individually offered a deal that if they betray the other, they will get off easier. That is if you betray your partner and he doesn't betray you back, you will get off free and he will get 3 years imprisonment. If he betrays you back, you both will get 2 years imprisonment. On the contrary, if you don't betray your partner while he betrays you, he will get off free and you will get 3 years imprisonment. The best outcome is achieved if you cooperate, if neither of you betray each other you will both get off with 1 year imprisonment. As you can see from the diagram below it is better if you cooperate. However, seen from an individual point of view it is always better to betray your partner regardless of what he does. Hence, they will most likely both betray each other and end up serving 2 years.

	B stays silent	B betrays A
A stays silent	Both serve 1 year	A serves 3 years B goes free
A betrays B	B serves 3 years A goes free	Both serve 2 years

Is the Prisoner's Dilemma relevant to the construction industry? Game theory suggests that any competitive situation involving a struggle for limited resources can be analysed as a game where players employ various strategies to achieve their goals. Traditional industry contracts often involve numerous parties, each pursuing individual goals and strategies, believing this will yield maximum benefits – possibly in the short term. However, renowned researcher John von Neumann demonstrated that cooperative players gain more in the game (Neumann, 2007). Thus, when examining the construction industry through the lens of game theory, it becomes evident that collaboration and win–win scenarios could pave the way forward for long-term success.

In win–win negotiations, the shared goal is to achieve a mutually beneficial outcome. This approach resembles the Prisoner's Dilemma, in which parties trust each other enough not to pursue a quick victory at the other's expense – a tactic that can often lead to extended penalties for both parties. Win–win bargaining operates on the principle that it is possible for both parties to attain their desired outcomes. This view is contrasted by zero-sum games or win–lose situations where one party's gain comes at the expense of the other.

In our pursuit of new cooperation methods, we strive to forge deeper connections with partners. Building close relationships requires trust, which can be challenging to establish and is based more on emotions than rationality. Key questions arise: do the involved parties know each other well enough for trust to be reliable? Is the project timeline sufficiently long for cultivating trust? Considering that future projects will introduce new players, the process must continually be renewed.

When construction project's profits start dwindling, it's tempting to abandon trust and prioritise personal gains. Most individuals understand this approach is unsustainable and negatively impacts the long-term trust 'account'. Trust is a dynamic value influenced by one's actions; prioritising self-interest undermines trust and prevents a win–win situation. Instead, the key to win–win success lies in collaboratively seeking solutions, being transparent about personal perspectives, and discovering alternative paths that preserve both project needs and bottom-line concerns.

In a mutually beneficial dialogue and collaboration, both parties work together to find solutions that help both parties satisfy their needs. This process requires open communication, mutual understanding, trust, and a willingness to consider different and alternative solutions than initially anticipated. By viewing the collaboration as an ongoing negotiation with two winners, both parties can achieve a more satisfactory outcome. You can basically follow this process to identify win opportunities in the dialogue:

(1) Agree that you will look for win–win opportunities together and discuss what could prevent it and what to do next.
(2) Build trust and invite the other party to do the same. Use active listening and foster understanding, honesty, and transparency.
(3) Ask questions and seek to understand the other party's perspective, goals, needs, and underlying interests. Ask what creates value for the other party. Pay particular attention to areas or solutions where both parties benefit from the cooperation.
(4) Be open about your goals, needs, and underlying interests. Talk about what creates added value or savings for your organisation. In the case of the Øresund Bridge, an earlier completion date was of value to the client. But it could also be other factors such as other deadlines, quality, publicity, risks, security, relationships, insurance, payment terms, and much more.
(5) Don't be overly quick with your choice of solution. Keep looking for creative solutions together that satisfy more needs – use brainstorming if necessary. The win–win principles should be reiterated as part of

discussions. If the value is primarily created by one party, it may make good business sense that the other party pays for half of the value created. For example, if the client has high liquidity and the contractor has low liquidity, there may be ways to change the payment terms for the contractor and the contractor can then give the client a discount equal to the contractor's savings on borrowing costs. If the contractor has another urgent job and the client is not in a hurry, the contractor can offer a discount for a later delivery, and so on.

(6) Make sure to write down all your win–win statements and their overall values in a common document. It may be relevant to document how the value of each factor is realised and documented to ensure a clear sense of fairness.

Create a safe and trusting relationship and be the first to share your thoughts and ideas with the other party first to allow for trust-building to take place. Then, ask the other party to do the same. In a win–win negotiation, both parties may have to give up something to reach an agreement. Be open to making promises that help both sides fulfil their needs and interests.

In the win–win mindset, both parties have a good feeling about decisions and commitment to the actions that have been agreed upon. When we think win–win, we leave the mindset of 'your way or my way', and instead, we take 'a third way', which is a joint path. Win–win situations are situations where both parties in a case are winners because they both gain advantage in the situation.

Win–win approach is based on an 'abundance mentality', where you think there is enough for everyone and are satisfied with your share and happy to share the abundance. This means you need to think more strategically and make long-term decisions. The opposite is a scarcity mentality, where everyone holds on to theirs, can never have enough, and doesn't believe there is enough for everyone. In his TEDTalk, Naveen Jain (2016) presents an exercise to illustrate the concept of an abundance mentality. If Jain should give one single advice to the audience, he told them to say out loud:

> I am going to dream so big that people think I'm crazy.
>
> I will never be afraid to fail.
>
> I will never think that anything is impossible because it will become impossible for me.
>
> I will have a mindset of abundance and I will never be thinking scarcity.

You can cultivate an abundance mentality in your everyday life by reflecting on your experiences and beliefs regarding your daily activities, projects, and the world around you. Focus on recognising opportunities instead of limitations and appreciate what you already have. Your beliefs influence your behaviour. Instead of thinking 'I never have enough' or 'I never do well enough', consider the idea that there's an abundance around you, with more than enough for everyone. We are back to the awareness of what is happening around you and your awareness of the opportunities that present themselves daily. Remember, what you focus on is what you attract – so if you concentrate on scarcity, things will seem scarce. On the contrary, if you emphasise abundance, you'll ultimately experience more of it. This notion may seem naïve, but it truly works – are you willing to give it a try?

Therefore, it's worth considering an alternative approach. Instead of forging ahead with a cooperation-based implementation, we should adopt a win-win mindset, allowing this conviction to shape our actions and behaviour in the partnership. This requires responsibility, honesty, trust, and a win–win mentality from all parties involved. Although it may be unrealistic to expect the entire construction industry to adopt this mindset simultaneously, some must initiate the change – and we can start by supporting others in our daily work. This transformation is a gradual process that demands courage, strength, and self-awareness. It requires stepping out of our comfort zones, exploring new paths, and discovering opportunities that others can follow and benefit from, much like open-source solutions.

For the prisoners in the Prisoner's Dilemma to succeed, it is essential for all parties to trust in cooperation and devise strategies that maximise benefits for everyone. To achieve this, we must foster new beliefs, strong leadership, systemic thinking, and shared visions. The presence of catalysts and change agents is crucial, as they can guide us along new paths and lead the way.

DESIGN OF PARTNERSHIPS

Partnering and strategic partnerships based on the win–win philosophy are the future of sustainable projects in the construction industry. It is the best way forward to solve the sustainability challenges of the future. The parties need to innovate together, utilise each other's competences and resources, and be creative together – creating synergy. Each actor plays a crucial role in ensuring the success of the whole, much like each Lego brick is essential in constructing a Lego figure. The true focus lies in delivering value for money and achieving sustainable solutions. By fostering better and more trusting partnerships, both efficiency and sustainability have the potential to significantly improve.

Tendering and contracting

Most partnerships begin with a tender. It is typically the developer who chooses the form of the tender, as well as the type of contract and the form of collaboration for both the consultant and the contractor contract. The alignment of expectations should start as early as the tendering and bidding phase. It is important that the developer already describes his visions for the cooperation in the tender materials, and thus what expectations the developer has for the upcoming collaboration partner. It should also be stated what the developer expects the tender will contribute and a description of what the developer's first proposal for the objectives of the partnership is.

Furthermore, the most important elements of the developer's organisation, used systems, and processes, as well as their approach to the execution of projects should be described. As far as possible, there should be an equality in the relationship between the customer and the supplier from the very beginning. This is only possible if the bidding party has precise descriptions of who the upcoming collaboration partner is, and what they expect from the collaboration. It also makes it much easier for the bidding party to make a proposal that matches the developer's wishes and expectations

The offers from the different actors must contain proposals for all partnership elements, and they are used as a basis to create the common frameworks with the developer and other actors within the agreement.

Partnership development

Establishing a partnership resembles constructing a small company together. Such a company possesses its distinct mission, vision, and values, along with the entire organisational structure. The partnership itself is cultivated collectively by all participants, taking into consideration the client's intentions and requirements.

The following list can serve as a useful checklist for developing a partnership:

- Organisation – roles and responsibilities and decision-making processes
- Financial matters and financial management
- A process for innovation
- Portfolio management – project types and complexity
- Joint skill development and activities
- Start-up and maintenance of the partnership

The last thing that should arise in a partnership is misunderstandings due to a lack of clarity about the above issues – both in the proposal phase and during the partnership. If the information is unclear, participants will resort to making assumptions or trying to interpret things themselves, which may not always align with the intentions of the partnership agreement.

Hence, it's beneficial to have a dedicated collaborative resource who can comprehend and support the growth of the partnership. Developing and executing joint initiatives demands resources, strength, insight, and persistence. Numerous details must be discussed, outlined, and implemented with complete transparency and accessibility on a shared IT platform.

A partnership developer can also streamline the process of establishing and onboarding partnership participants within the joint organisation. It is essential for this process to be well-defined and prepared from the outset of the collaboration.

Organisational set-up

The partnership creates an organisational structure that encompasses:

- Description and establishment of common organisation, roles and responsibilities, and key decision-making processes (governance structure).
- Meeting structure for the steering group, portfolio management, operational management, project management teams, and so on.
- Common work processes for participants.

When establishing the organisation, it's crucial to clearly define all roles within the partnership and their respective responsibilities – like organising individual projects. Essential communication and decision-making channels should be explicitly outlined and made accessible to everyone involved. This fosters a sense of certainty and trust, ensuring that all parties within the coalition comprehend the process for meetings and communication among partners, project participants, steering groups, and stakeholders right from the beginning. It's vital to prevent any confusion regarding decision making in daily operations and project work.

Establishing routines early on is both easier and more effective than implementing them belatedly – be proactive. For instance, projects will inevitably undergo changes, and having clear processes in place ensures a sense of psychological safety for all parties involved, particularly when it comes to decision making and subsequent communication within the organisation. Other process descriptions may address innovation, evaluation, knowledge sharing, commissioning, and so on.

Most agreements incorporate a mandatory, systematic approach to proactive conflict management. The parties involved will collectively establish detailed procedures for resolving conflicts, preferably documented during peacetime. Furthermore, having a conflict resolution strategy in place has proven to reduce the occurrence of conflicts, and those that do arise are resolved more swiftly and with minimal complications. Effectively managing interface risks and responsibilities is also very helpful in terms of eliminating potential conflicts. The more conflicts that are prevented early on, the better. After all, projects regularly encounter minor and major conflicts throughout their progress.

Financial matters and financial management

The economic model must describe all economic processes and financing relationships.

Nothing should be left to the imagination, as disagreements about finances are a breeding ground for conflict. The financial model for each project must be drawn up in such a way that it is clear to all parties what is included and what the underlying costs are. It must be transparent and clear to everyone, otherwise things will go wrong.

The agreement may include various incentives or bonuses and a model for distributing any surplus or deficit fairly between participants according to the volume of work or value created. The financial models help to ensure that everyone is working with the same, or at least fairly aligned, financial incentives. This minimises the problem of the recurring questions about extra work and extra bills. When extra work and extra costs are required, they will be actualised. In all other cases, everyone shares an interest in getting things done at the lowest possible cost. This reduces or eliminates the mistrust that may otherwise easily arise about the lack of financial fair play between client and contractor.

Innovation

To maximise value for money, joint learning and development should be integrated to meet overall objectives and specific project goals, ensuring optimal solutions. Establish processes, boards, and development sessions to identify and implement small continuous improvements and to trial and evaluate innovation elements benefiting one or more projects. A clear innovation strategy is essential, outlining where, how, and when innovations will be realised, as well as who will incorporate them into the projects. Moreover, there must be clarity on the management of risks and interfaces within and between projects – specifying how, when, and who will handle them.

Portfolio management of projects

Project portfolio management is a crucial tool within the partnership, serving as the foundation for initiating and implementing projects that create value for all involved parties. It facilitates the prioritisation of projects based on their launch timing. The partnership's success depends on whether portfolio management can help optimise resources, reuse lessons learned, and create a flow in the project delivery plan that satisfies all stakeholders. Some projects will take longer or use more resources than expected, or vice versa, and a portfolio management model must be able to accommodate for this.

Portfolio management ought to be managed in a specific meeting by a specific organisational unit. There must be clarity on the criteria underlying the prioritisation criteria that determine the launch of projects in the partnership. Systematic management and periodic reviews of the project portfolio should help to determine the most optimal portfolio for all actors at any given time, even if it may be necessary to postpone or accelerate projects. Portfolio management also needs to be managed if some projects require more focus than others.

Joint skill development and activities

In collaborative partnerships, it is beneficial to engage in joint skill development in areas like teamwork, strategic planning, health and safety, sustainability, and conflict resolution.

Shared facilities, such as co-working spaces, can significantly enhance collaboration. When team members are in close proximity, they tend to communicate more effectively and resolve issues as they arise. Although meetings serve a similar purpose, having a shared workspace – even part-time – greatly benefits all parties involved.

In addition, team-building activities during onboarding and throughout the project can further foster cooperation. Socialising and understanding each other on a personal level proves to be beneficial.

Below is a list of common activities (not exhaustive) that can be implemented in a partnership agreement:

- Start-up workshop for all participants.
- Strategy workshop for the management of the partnership.
- Idea generation workshop (at the beginning of the project).
- Workshop – development of project culture including values, common rules, conflict management ring, and so on.
- Design/planning meetings and workshops.

- Phase evaluation – what went well and what do we need to improve in the next phase?
- Follow-up meetings – partnership development.
- Ongoing communication of partnership standards and processes/value streams to the field work processes.
- Follow-up/evaluation workshops focusing on development initiatives/ work processes/joint organisation/cooperation and so on.
- Joint training on partnership – change, communication, project management, partner closeness of the relationship.
- Preparation of process plans (lean construction).
- Weekly meetings with design management and construction management.
- Board meetings on the construction site.
- Newsletters.
- Establishment of an innovation board (ongoing discussions on different initiatives for
- process product).
- Identifying and communicating the 'good story'.
- Common reporting routines to the steering group.
- Sparring with project participants.
- Co-creation activities.
- Mediation.

The importance of communication in projects is frequently overlooked, so it is essential to focus on incorporating communication elements within the partnership. Be mindful of aspects such as scope, format, and content.

Start-up and maintenance of the partnership

Partnerships may consist of large teams, and they require clear communication. All partners in the partnership should sit down together in the start-up phase and have a dialogue about the expectations and goals of participating in the partnership. Everyone ought to, if possible, be involved in developing common objectives and collaborative intentions and values. These should be written down and communicated massively through all the channels established in the project, and there should be a continuous follow-up on whether the intention is honoured. The intention of cooperation should foster the desired partnership culture. It must be cultivated and grown through confidence-building activities and dialogue, dialogue, dialogue.

The onboarding process focuses on getting new partnership staff members ready to join the partnership as quickly as possible by introducing them to the organisational set-up, portfolio management of the projects, and the various facilities of the partnership itself. As part of the onboarding process, participants at all levels contribute to establishing and documenting the meeting structure, work processes, and joint activities.

Ongoing evaluation of the project cooperation and the participants' performance must be initiated early on, even if things are going well. Evaluation should be carried out on an ongoing basis and at least at major milestones. This ensures that positive and negative experiences are captured, and that elements of the partnership and aspects of project implementation are adjusted on an ongoing basis. Furthermore, it focuses attention on the future of the project and especially where improvements can be made regarding work processes, organisation, project execution, or other areas.

MAIN POINTS

It is time to explore partnership methods to ensure optimal conditions for future projects. We ought to demonstrate how this kind of collaboration can lead to measurable results in sustainability, human satisfaction, enhanced working relationships, and a balanced approach to time, cost, and quality. We need to develop to move away from the competition-driven focus on being the fastest, best, and cheapest that currently dominates the industry. Thankfully, the partnering mindset is gaining traction and offers solutions to many industry challenges. However, it is crucial that partnering is implemented thoroughly, not just superficially.

Long-term strategic partnerships are gaining ground, and that's a good thing. Under the shelter of partnership agreements, there is time to create a common approach to delivering projects and achieve far better results than if projects were delivered individually. Strategic partnerships create joint development opportunities and can help lift everyone in the group to a new level. Such partnerships offer joint development. A lot of work is involved in developing partnerships in the right direction. All parties to an agreement must be open about their own strengths and weaknesses; a fair partnership must be created from the start, and development can start from there. It is an investment where the payoffs will only emerge over time. But do we have a choice? The future is created together.

The win–win philosophy can be successfully implemented if we all adopt the mindset that the world is abundant with opportunities, and there is enough for everyone. Imagine if more people operated with trust and integrity. Our focus should be on treating others in the industry as we would want to be treated ourselves.

Initiating change requires courage and thoughtful consideration. If you possess the courage to begin, what's holding you back?

Embracing partnerships can be challenging at first, as it goes counter to our basic instincts. We must adopt the belief that there is enough for everyone and recognise that, at times, a smaller portion of a bigger cake is better than a larger part of a smaller cake. Additionally, recognise when you have had enough.

YOUR REFLECTIONS

What was new to you in this chapter?

Did anything surprise you or capture your attention after reading the chapter? Did you identify any elements that could be beneficially implemented in the project? Who should you involve initiating the process? Are there any people around you who have prior experience and can help?

In which ways and to what degree have you experienced partnering and/or lean construction?

Reflect on your past projects and involvement in collaborative efforts. Can you recall a particular successful experience from a partnership project? If you had to pinpoint one factor that contributed to its success, what would it be? Do you continue to apply the lessons you've learned, either by directly using them or by adapting them to new contexts? How did you utilise planning tools during the collaboration process?

Could you share your experiences with lean construction application? If given the opportunity to enhance lean construction, what modifications would you suggest?

Have you ever experienced or participated in a partnership agreement?
Can you recall your fondest memories from such collaborations? What role did you play, and how successful was the outcome? Would you consider doing it again and why?

Shaping the future together
What are your thoughts on building the future collectively? How can you contribute to this endeavour? Are there people around you who might be interested in forging new forms of cooperation? What is holding you back? What could get you started? What are you missing? Someone has to start – is it you?

READ MORE

https://eci-online.org/publications/ contain a number of relevant publications from ECI – the European Construction Institute (ECI, 2023).

The book 'Spiral Dynamics' describes a new framework for understanding the dynamic forces at work in human development and interaction. It is about how we evolve through a series of equilibrium states, based on the theories of Clare Graves (American psychologist). Each level of 'value state' has a new specific view of the world, through its specific filters. Each of the levels can be very different, and so can the decisions made there (Beck, 2014).

In the book 'Radical Collaboration' (Tamm, 2005), the author provide tools for creating better collaborations and becoming more aware of other people in the collaborative relationship. The book is also useful as a guide for those who want to create trusting environments for collaboration.

In the book 'Getting to a Yes – Negotiating without Giving In', the author focuses on the dialogue and the process in actions where the starting point is to create a windfall (Ury, 2011).

doi: 10.2166/9781789063349_0215

Chapter 9
Co-creation

Don't say it can't be done, say instead that you don't know how to do it yet.

Mr. Tomas Bata

Co-creation is a process aiming to create something new and valuable in a joint endeavour. What is the difference between co-creation and collaboration many ask? The difference is that collaboration is a broad word that includes many different types of purposes. Co-creation is a newer word with a sharper focus on creating something new of value together. Some well-known co-creation methods fall under the innovation category, such as 'open innovation', which welcomes contributions from external stakeholders; 'crowd sourcing', where challenges are open for public input; and 'design thinking', which involves community-driven prototype development.

Co-creation is characterised by cooperation, where the participants all take an active part in the process. Everyone has a shared responsibility for the outcome, and the aim is for everyone to benefit from the outcome. Creativity and trust are key qualities of the cooperation which is a prerequisite for co-creation processes to work. Furthermore, a certain level of flexibility and adaptability is essential among all parties, ensuring that no one clings too strongly to their own ideas, allowing the joint process to progress.

This book on 'Sustainable Project Management' is an example of a co-creation process in which we – two people with different experiences – collaborate to answer the question 'how can sustainable project management be part of how we create the future's sustainable society?' The process cannot be fully planned in advance because the participants do not really know from the start what they are creating. It is only through the co-creation processes that this becomes clear. Consequently, ongoing discussions about the process and next steps are frequent. A combination of various process methods is often used to succeed. The atmosphere during the process needs to bear the characteristics of psychological safety. Psychological safety ensures that every participant has the needed courage to 'go out there' as well as access to their inner resources in terms of creativity and imagination. This is the fertile ground for something new to emerge.

If you aspire to foster more co-creation as a manager within your organisation or project, consider adopting promoting co-creation, engaging team members in decision making, and maintaining openness during both successes and challenges. Additionally, provide opportunities for creative problem-solving exercises, facilitate co-creation processes, and make a conscious effort to define your objectives clearly but leave the decision about the way to the reach the objectives to the participants.

This chapter outlines a few methods for co-creation. The complexity of the methods increases throughout the chapter. As a project manager, it is advisable to begin with a level you feel comfortable with and gradually enhance both your own skills and the group's ability to co-create. We suggest adopting an experimental approach and then venture out of your comfort zone one step at a time.

Some basic principles about processes that work are:

- Make the purpose clear.
- Ensure a diverse and inclusive approach.
- Facilitation is key – make it easy for everyone to contribute.
- Everyone should contribute and, ideally, take 'something home with them'.
- Often, a slower pace is necessary compared to the speed at which most people usually work – think tip: we start out slow and then we decrease the speed.

Many meetings encompass a variety of facilitation methods. Specialised facilitators may be able to skilfully blend and adapt methods to ensuring high-quality processes – and it can be a joy observing the real experts. Project managers often and inherently play a facilitative role in their work, so we need to have some kind of basic understanding of these methods. Inadequate facilitation can result in projects that become too narrowly focused, as a limited number of individuals monopolise the process and prevent others from utilising their skills. This can also create a more stressful environment, with fewer individuals taking charge and contributing. Insufficient co-creation may lead to inappropriate project outcomes and disheartened participants who feel their opinions are inadequately valued. Therefore, it is a good idea to understand and practice co-creation methods – only through practice it can get better.

THE GOOD MEETING

Several things can go wrong in meetings; hence, good meeting management is in high demand in many organisations. Project managers often have the role of chairing meetings and therefore needs the basic understanding of what it takes to make meetings work.

Checklist to prepare and ensure a good meeting includes:

- A clear purpose of the meeting is described.
- The purpose, agenda, and materials are sent out in advance, preferably well in advance.

- The expected output from the meeting is clarified as far as possible.
- The correct participants have been invited and have indicated their intention to attend.
- The necessary time for the meeting has been allocated.
- The format of the meeting has been decided and the venue has been finalised.

To ensure effective meetings, it's important to avoid certain behaviours. The meeting leader can greatly enhance the quality of the session by adhering to specific criteria:

- Speak slowly and clearly.
- Begin the meeting with a minute of silence, allowing everyone to adapt to the meeting environment.
- Be cautious with 'funny remarks' – they can easily be a source of misunderstanding.
- Try to anticipate if conflicts may arise in the meeting, there may be more disagreement and different interests than it looks on the surface.
- Create an overview, utilising drawings and illustrations to clarify your points when appropriate. Point and explain.
- Utilise the purpose and agenda effectively during the meeting to ensure unrelated topics are set aside.
- 'Park' unresolved issues for later to prevent them from consuming all the available time from other important issues.
- Ensure a balance between being well-prepared and clear, while remaining open and relaxed to be receptive for questions, concerns, and perspectives that are new to you. In the best meetings, everyone gains new insights and learn.
- Allow time for questions and discussion.
- Summarise as you go along.

Patience is important as a meeting facilitator – the bigger the meetings, the more patience you need. You may find some questions tiny or irrelevant, or that they are about things that were agreed upon long ago or details you don't know the answer to. As a moderator, it is your job to keep calm, be friendly and polite, and ensure that all participants can take part regardless of their starting point. Put away the kind of pride where you feel you need to know most or be the best. Instead, admit openly when you don't know and ask back. Asking questions to gain insight into unfamiliar issues fosters trust in the group.

If you find meeting participants becoming 'difficult', irritation may creep in under your skin. However, keep calm and try to focus your energy on understanding their perspectives, enquiries, or attitudes. If the facilitator of a meeting is inattentive and fails to comprehend participants' feelings and reactions, small conflicts may arise, hindering the meeting's desired outcome to materialise. Therefore, try to listen actively and ask questions. Be careful about the ideas and motives you attribute to others. Remember that you know less than you think. You rarely know how the world and the project look seen through other people's eyes. So, be gentle and considerate.

In general, it is therefore important that the moderator tries to avoid:

- Low-calorie talk – make sure messages in your speech bear meaning and significance.
- Treat everyone with respect and as equals, regardless of gender, role, style, type, and so on.
- Address objections – strive to comprehend the concerns and underlying needs.
- Avoid confrontations and conflicts – utilise de-escalating language.
- Refrain from 'snarky' remarks – maintain focus on the project and meeting content.
- Conclude meetings by summarising decisions, key conclusions, or insights. Consider documenting complex decisions in a shared document or flipchart visible to all participants. This approach fosters collaboration in finalising the decision.

Facilitation

Facilitation comes from the Latin word 'facilities', which means to make something easier or smoother. Facilitation has become very popular in the last 10–20 years. Facilitation helps to improve communication and collaboration. Facilitation allows all participants to focus on the content while the facilitator ensures a good process.

In facilitation, the facilitator assists a group in achieving their goal through a group work and a dialogue-driven process. The responsibility of the facilitator is to support the group along the way steering the process. The facilitator should strive to remain impartial regarding the final outcome, focusing on guiding participants towards discovering the best solution on their own. Facilitation is not suitable for situations where decisions have already been made and the meeting's purpose is to inform or seek acceptance of a decision – this is a kind of dishonest way of misusing facilitation. So, apply facilitation when you genuinely want people to contribute with their ideas.

Facilitation is effective in situations such as:

- Aligning expectations – when the group needs to reach a common understanding of a topic.
- Project initiation – where project members collaborate on knowledge sharing and planning. The five essential tools can be employed for organisation.
- Decision-making meetings – where a decision must be made, a solution selected, and further actions determined.
- Strategy development – where an overall direction needs to be discussed and decided upon.
- Innovation – generating new ideas and solutions for the project or problems in the project.
- Evaluation – reflecting on the progress and work completed thus far.

The facilitators' job is to tailor the process according to the purpose of the situation or meeting. They prepare the process to assist and support participants before, during, and after, ensuring real results of the collaboration. As a project manager tasked with facilitating and designing a process, there are several factors that need to be considered:

(1) Formulation of purpose and expected outcome.
(2) Decide who participates and in what capacity.
(3) Defining the process and methodology.
(4) Outlining the further process after the session.
(5) Creating the appropriate space, environment, and atmosphere.
(6) Role allocation – who does what during the session.

The most crucial aspect is to define the process's purpose, expected outcome, and whether a tangible deliverable will result from the session. Carefully choose the appropriate participants who can contribute with their expertise in the group while taking into account the dynamics among participants.

It's essential to design the form and methods to optimise desired outcomes. These outcomes are often finalised and documented at the end of the session. The process typically involves a mix of group work, reporting to the plenary, joint dialogues, documentation, and breaks for reflection.

Make sure you prepare materials to support the process beforehand. Make sure that the material is as unambiguous as possible. In a group, there is always a wide range of possibilities for interpretations of instructions, and it is not always easy to ensure that the end-product is useful.

A crucial task for the facilitator, both before and during the event, is to ensure a positive environment and atmosphere. The room's layout is more significant than one might realise, as is the choice of location and seating arrangement. These seemingly simple and tangible factors greatly influence the group's dynamics. Additionally, assigning roles is essential. Determine who is responsible for which tasks, involve participants in the process, and establish clear meeting rules.

It is beneficial for the facilitator to prepare the entire process in a well-organised script. This script should outline each aspect, including the purpose, activity, responsible person, materials, and arrangement. The script also serves as a useful to-do list. For larger sessions, having all details written down ensures that nothing is overlooked. It's crucial not to underestimate the importance of preliminary tasks such as logistics, catering, table arrangements, registrations, room set-up, and securing necessary materials prior to the session's commencement.

The facilitator must be prepared to adapt the process if the planned elements don't work as intended. If a change is needed, ensure that the consequences are explained. The facilitator's primary focus should be on the process and meeting the participants' needs and engagement.

Here are a few facilitation tips:

- In the introduction, take advantage of one of the best opportunities to make participants feel welcome and emphasise the importance of their participation.
- The introduction should include the purpose, programme, presentation of participants, and materials to be used, at a minimum.

- Explain the process layout, expectations for each participant, and provide an example to illustrate the method.
- Apply active listening to demonstrate your interest in seeing, hearing, and understanding all participants.
- Monitor everyone's contributions and ensure they are heard, while remaining aware if parts of the topic are not understood. Highlight participants' solutions so that everyone can see the potential.
- Encourage participants to provide input, listen actively, give space to others, and make efficient use of time.
- Periodically summarise interim results.
- Manage time in a considerate yet precise manner. Clarifying time frames and explaining why a discussion is stopped (often to accommodate other discussions) can be helpful.
- Conclude by summarising the meeting and communicating the next steps.

Use your voice, body language, and senses to create an atmosphere of easy, safety, and energy. Don't be a 'know-it-all'. As a facilitator, it's your responsibility to ensure that the session's outcomes are highlighted and documented. Becoming a skilled facilitator requires training, but it's an achievable craft through practice and preparation.

MEDIATION

Mediation is traditionally a tool for the peaceful (i.e., non-violent) resolution of conflicts. However, the mediation is also applicable to problem solving, and the underlying principles are very useful for group facilitation.

Most people are so eager to find solutions that they tend to find solutions before they even understand the problem. Or said in a different way: they understand the problem but only from their own perspective. When they subsequently present their solution, their solution comes into conflict with other parts of the problem that they have either not understood or underestimated. This is fertile ground for conflicts. Thus, a mediator's (think facilitator) role is to ensure that everyone comprehends the problem seen from multiple perspectives before starting to generate solutions. In successful mediation processes, groups often spend about 80% of their time understanding the problem and only 20% finding solutions(!) Once the issue is clearly understood, creating solutions becomes significantly easier. Better understanding, that is, from more perspectives and in greater depth, leads to better solutions. This results in solutions that work for more people and hence are richer in the qualities related to sustainability.

The mediation process consists of five phases and is based on the principle of identifying needs and interests as a starting point for understanding the problem (Monberg, 2019). The problem doesn't need to be conflict in the usual sense of the word but can be viewed as a collection of needs around a topic. Many people struggle to convey their underlying needs; often they are not even clearly aware of their needs. Hence, recognising needs often necessitates training or facilitation. Some people have a tendency to express their needs as hard demands. This is generally not very conducive to finding joint solutions. The demands often come from underlying needs that can be met in many different ways and not only be meeting the initial hard demand.

A banal example: a child proclaims a hard demand for a soft drink may now! This demand may cover over the need to satisfy thirst – or the need for comfort – or the need for excitement and energy. When you discover the underlying need there may be better solutions than soft drinks.

Particularly in conflict situations, the stories of the conflicting parties are often two positions – almost like two bastions from which the parties shoot at each other and each other's stories. Thus, there is often a preparatory phase (phase 0), preceding the five phases of the mediation process. This initial step involves gathering all parties around the table with a shared intention to trustingly engage in the process. This often includes creating a safe atmosphere for all, discussing the process and making sure that rules of confidentiality and kindness are understood and accepted by all. Creating a good atmosphere from the beginning is very important.

The five phases in the mediation process are summarised in Table 8.

Phase 1: Storytelling

This phase is often the most extensive, and it is crucial to allow time for diverse perspectives to be heard (80%). Active listening is essential for everyone to feel heard, and it is important to consider all perspectives. As everyone in the room will participate in decision making during subsequent phases, understanding the background of the solution space and the criterion for a good solution is necessary.

In the exploration of individual stories, it is useful to recognise that each individual's story resembles an onion comprised of layers. Within these layers, vital information can be found (see Figure 26).

Table 8 Five phases of the mediation process.

Phase	Name	Purpose of Phase
1	Storytelling	Everyone tells their own version of the story and thereby the story is understood from different perspectives.
2	Identifying needs	Joint brainstorm on the needs, concerns, and interests of all parties ensures needs are identified and everybody are put in contact with their inherent empathy.
3	Proposed solutions	Brainstorm solutions that resolve parts of or the whole conflict – at this stage no decisions are taken – only looking at possible options.
4	Design of solution	From the ideas in phase 3, an overall solution is chosen. A description of a future that all parties live with and see themselves in is prepared together.
5	Conclusion of the agreement	An agreement is drawn up on how and when the agreement will be implemented and who is responsible for what action points.

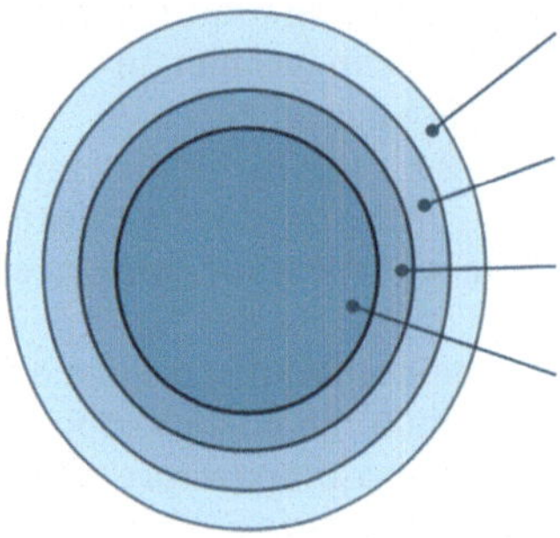

Figure 26 All stories telling about real events have different layers. It is a good idea to understand the different layers of the story if you want to understand needs and find a solution in mediation.

The outer shell of the story contains facts, events, and claims and is straightforward. This section is one that most people can recite effortlessly, as it typically receives the most attention when the story is told. This layer helps ensure a common understanding of the overall sequence of events. In complex cases, consider creating a timeline of events and including the key points that all parties agree upon. This approach streamlines discussions and helps everyone navigate the situation more easily.

The next two zones are about the beliefs and emotions involved in the story. Opening this layer of the story requires building trust in the mediator and their ability to manage the group. It often involves handling individuals who struggle to stay within the context. Nevertheless, many people wish to express their inner thoughts regarding the conflict. It's not about 'revealing one's emotional life'; instead, one can utilise questions such as:

- 'What do you think about what happened/what you have heard?'
- 'What idea do you have about [the incident or motives] ...?'
- 'Has what you've heard here altered your perception of the situation or incident in question?'
- 'What do you think about what happened?'
- 'What does it make/made you feel?'
- 'What did it do to you?'
- 'How do you feel about ...?'
- 'How did you react to ...?'

A key insight to consider is that people often judge themselves by their intentions and others by their actions. This means that if someone feels wronged and confronts us, we may think it's unfair because our intention was different from how they perceived it. On the contrary, when we encounter inappropriate behaviours in others, we tend to judge based on their actions, finding it difficult to see or understand their underlying intentions. It's particularly challenging to believe that someone had a different intention than causing us harm when we feel hurt. This very mechanism can obstruct reconciliation and active listening.

Keep this notion in mind: 'People judge themselves by their intention and others by their action'. Thus, a central goal of mediation is to reveal each person's true intentions, emotions, and actions – that may have failed. When one party recognises this in the other, anger often dissipates, creating space for collaboration towards a solution. Something special happens in the room when one party realises this about their counterpart – it is as if the anger melts away, and it is possible to come together to find a solution.

Once mediators believe the situation has been thoroughly examined from multiple perspectives, it's time to proceed.

Phase 2: Identifying needs

We can all support each other in identifying needs, concerns, and interests. As individuals, we sometimes hesitate to express our needs, perhaps due to the fear of appearing 'needy'. Therefore, it is beneficial to encourage all parties involved to try and understand and communicate each other's needs. Rather than solely focusing on sharing your own needs, it is important to actively listen and clarify your understanding of other's needs. We must recognise that what we consider reasonable and obvious needs ourselves may not always be so apparent to others. Additionally, it is crucial to acknowledge that needs can encompass various aspects, including physical, financial, and emotional requirements, such as the need to be heard, consulted, involved in decision making, and recognised. Exploring Maslow's hierarchy of needs can provide valuable insights into the different levels of individual needs as shown in Figure 27.

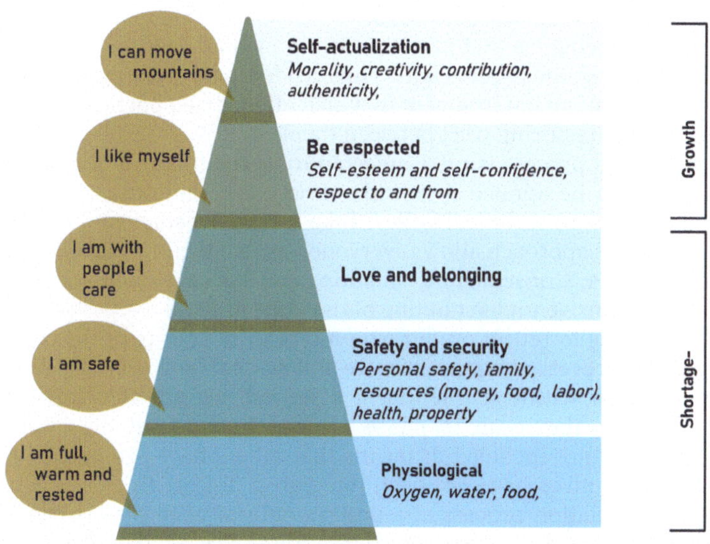

Figure 27 Maslow's hierarchy of needs symbolises that people first and foremost need to have needs met at the bottom of the pyramid; only as they are adequately met do needs emerge higher up the pyramid (Maslow, 2012).

Phase 3: Finding solutions

Finding solutions is surprisingly often the most straightforward part of the process. Once we comprehend the problem from various perspectives in phase 1, and recognise specific needs, concerns, and interests in phase 2, it becomes effortless to pinpoint areas where we can negotiate or address significant needs of the other party with minimal investment. Collaborative actions or projects might also serve as viable solutions. The suggested ideas are typically produced during a brainstorming session without extensive discussion or evaluation of their merit.

Phase 4: Design of the solution

In phase 4, you collaboratively review the solutions proposed in phase 3 and select the solution(s) to work on together. This solution should be balanced and mutually agreeable, ensuring that everyone feels satisfied with the chosen joint solution. Ideally, the selected solution leads to a complete resolution of the conflict – but even a partial solution can be helpful.

Phase 5: Conclusion of the contract

It may be helpful to document the solution in writing to ensure that everyone remembers the agreement and to properly allocate costs when relevant. Additionally, creating a timetable that outlines specific actions and responsibilities is crucial to ensure the successful implementation of the solution and to clarify the roles and accountability of each person involved. Although in some cases a verbal agreement may be sufficient, it is important to determine this on a case-by-case basis.

It is vital to recognise that participation in mediation is voluntary; without this understanding, the process fails. Thus, aside from resolving existing issues, the goal is to foster an environment in which all involved parties can coexist in harmony, ideally restoring trust between them.

The mediation process is often referred to as the 'natural decision-making process' and can be applied in a wide variety of situations, including those involving complex or technical issues. Take it stage-by-stage and not jump too fast forward. This approach allows everyone to voice their thoughts and include their perspectives. Consequently, solutions can be developed collaboratively with a comprehensive understanding of the issue at hand.

It is interesting to reflect on our own internal thought patterns considering the mediation process. In principle, we follow the same pattern, but we do it so quickly that within a fraction of a second, we are able to convey the final outcome as the result in phase 4. Consequently, all the intermediate deliberations are not included. It means that others have not been consulted, and their perspectives have not been integrated. Taking the time to progress through the mediation process at a group's natural pace proves beneficial in resolving numerous issues that could have otherwise escalated into conflicts. Instead, everyone can depart from the mediation process with clearly defined roles and responsibilities, along with a deep understanding of the reason behind each of the delegated actions.

Preparation for the mediation process can be carried out step-by-step:

(1) Consider how the parties and the mediator should be seated.
(2) Decide on a start-up script, which should include:

   (a) Description of what is going to happen in review of the process.
   (b) State which parts of the process are 'set in stone', those that can only be influenced, and those that offer more flexibility?
   (c) What is your overall hope for the process. For example:

      (i) We have a situation that is not working; we must try to find a solution that works for all/both parties.
      (ii) That resolving the conflict means (e.g., peace, innovation, cooperation, and so on).

   (d) Consider if you expect resistance from some persons and try to address their concerns by proactively articulating the resistance and how you have thought about how to deal with it. (E.g., 'I know this conflict has been going on for a while and that it can be a challenge to be here today. So, I would like to recognise you all for giving it a try today'.)

(3) Allocate time prior to the mediation to centre yourself. (Some quiet time alone can be helpful.)
(4) Remind yourself that whether the mediation is successful or not, you will not blame yourself. Openness, caring, imagination, and honesty are the golden standard, so put away any sense of 'performance' thinking.
(5) Consider power dynamics between parties: it is crucial to comprehend whether the parties involved are on equal footing or if there's a power imbalance. This could be attributed to factors such as experience, rank, emotions, fears, potential threats, communication skills, authority, and knowledge.
(6) Consider the empty chairs: there are almost always parties to the conflict who are not present. You should ensure fairness for them as well.

There may be instances where parties feel uneasy during mediation, and it's the mediator's responsibility to help them feel safe in the process and in the framework of the mediation (see Figure 28). This can be achieved, for example, by maintaining a record of any concerns and establishing ground rules to minimise discomfort.

Questions that can help parties into the framework:

- 'What could you gain from being here today?'
- 'What would it take for you to be ok with being here today?'
- 'In what way is it difficult for you to be here?'
- 'If a solution is not found today, what would your situation be?'
- 'In that case, what would you do next?'

It is possible to strengthen the framework by asking: 'What is important to you about being here today?' The more people present, the more often the framework should be repeated.

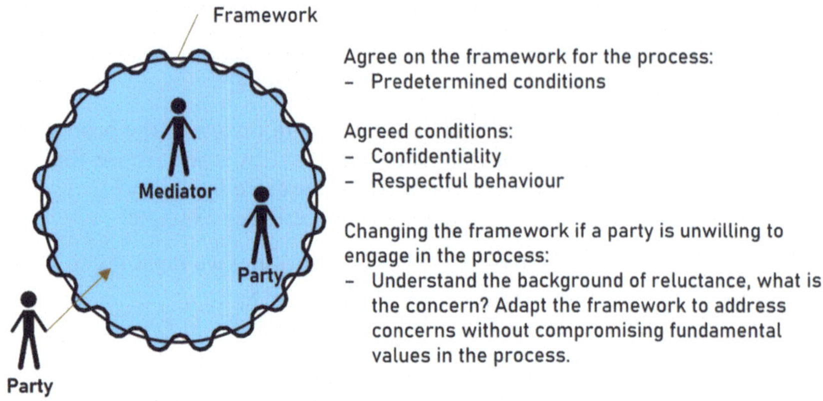

Figure 28 Framework for the mediation; what has been agreed and what should happen if the framework changes.

To create a setting where everyone feels respected, the Donna Hicks dignity elements can be a good reminder. Donna Hicks' elements of dignity help us all to remember what it takes for everyone to be treated with dignity (Hicks, 2021):

(1) Acceptance of identity – meet X as neither superior nor inferior regardless of gender, race, religion, position, and so on.
(2) Inclusion – making X feel welcome and part of the group.
(3) Safety – making X feel safe to speak without causing retaliation.
(4) Acknowledgement – give X your full attention and acknowledge worries, feelings, and experiences.
(5) Recognition – value X for his/her ideas and contributions.
(6) Fairness – be fair and non-discriminatory.
(7) Benefit of the doubt – start with trust and faith that X acts with integrity.
(8) Be understood – what X thinks is important, give X the chance to explain his/her point of view.
(9) Independence – encourage people to act according to their own convictions and thus control their own lives.
(10) Accountability – take responsibility for your actions and if you have offended X's dignity, apologise.

SEE THE WORLD AS A FUTURIST

As a project manager, it's essential to consider not only the project's lifespan but also the long-term impact of the product it creates. In the water industry, the result of many projects may still stand 30–100 years from now. Keeping such an extended time horizon in mind can be challenging. Unfortunately, we often become preoccupied with immediate concerns and priorities, hindering our ability to think things through properly. If your team can successfully create products that future users will appreciate the project can be said to be truly successful.

A crucial aspect of management is ensuring a shared vision and mutual understanding of the future. In the past, this was determined by the person at the top of the hierarchy. However, with modern tools, collective collaboration is now enabling anyone to be an 'amateur futurist'. Our brains naturally act as futurists, constantly contemplating and preparing for what lies ahead based on past experiences. By utilising appropriate tools and working together, we can derive insights about the future from varied perspectives beyond our own experiences. This collaborative effort will result in clearer, more informed visions to guide our decision making.

As a futurist, it's not about being right or wrong. Not even the professional futurists can accurately predict the future. However, if we recognise that it is about understanding the world and contemplating its direction and prevailing trends, we can derive inspiration from these insights to develop incredible solutions that may help shape the future.

In their book 'Awaken Your Future Sense', Anne Skare Nielsen and Henrik Hovgaard (Hovgaard, 2020) write about the paradigm shift from industrial to sustainable. They describe the problem as a game metaphor and write that you can play the 'future game' in two modes: tribal or planetary mode.

When you operate in 'tribal mode', your primary concern is: 'What's in it for me'. One can both see and hear this; you talk a lot during meetings and don't listen much to others. You often talk about how you want things and give orders left and right. The focus is always on speed and the lowest price. You are keen on the management recognising your work because you would like a raise, a promotion, and so on.

However, when you approach the future in the 'planetary mode', your focus is on collaboration. This means that you focus on making the project work in a way where the work is distributed evenly among everyone, and everyone's input is valued and turned into worthful solutions. You manage to focus both on the task and on the people in the project. You have visions for the project and want it to contribute to a good future. You are aware of what you are doing, and you take the breaks necessary to make wise decisions. If there's something you don't know, you ask others for advice and find knowledge both within the team and from the surroundings. In the 'planetary mode', steering committee meetings are not an opportunity to show off how skilled you are but an occasion to become clearer on the vision and principles. Instead of getting answers to specific questions, you are more interested in reaching an agreement on good, healthy, and efficient principles, which you and the team can use to navigate independently. In the planetary mode, you get a thrill from facilitating the team to find new innovative solutions that truly make a difference for a sustainable future.

In 'Awaken Your Future Sense', the authors quote the old Apple advert from 1997: 'Here's to the crazy ones. The misfits. The rebels. The troublemakers. The round pegs in the square holes. The ones who see things differently. They're not fond of rules. And they have no respect for the status quo. You can quote them, disagree with them, glorify or vilify them. About the only thing you can't do is ignore them. Because they change things. They push the human race forward. And while some may see them as the crazy ones, we see genius. Because the people who are crazy enough to think they can change the world, are the ones who do'.

With this reference, they challenge us to think rebelliously about the future: what could the future bring if we aspired to be one of those 'crazy ones'?

When we use our brain's ability to predict without reflection, we tend to base our thinking on past experiences. In this way, we recreate the past. To do something new and better aligned with the future, we need to imagine the future through a different and conscious process. A simple and rather central observation in Skare and Hovgaard's book is that the changes that are going to happen will happen through a new kind of competition. A competition for the best ideas about what the future should look like. It is on 'visions of the future' that the battle of the future will be fought.

To do future research is an effective collaborative exercise that enables a team to envision their ideal future together. By doing this, they can identify new opportunities. As they assess the project's future, they collaboratively develop a shared vision of the future they aim to achieve.

As a concept for effectively engaging with the future, futurists analyse trends. Trends represent significant shifts from one state to another, possessing transformative power that can alter the way we work, think, prioritise, and make decisions. These developments hold the potential to revolutionise our approach to various aspects of life.

You can choose to work with one or more trends. It is effective to focus on one or a few game changers. What is a game changer? It's a trend that you have a hunch will change the rules for how you and the project should work to succeed in the future. Yes, be a co-creator of the future.

All trends have the structure from A to B. Skare and Hovgaard argue that perhaps the key trend affecting all other trends these years is the trend 'from more to better'. How might this trend appear in the context of water projects? What does it take to prioritise quality and efficiency over quantity? Which aspects of value creation should be the main focus?

If we consider the importance of creating higher-quality homes rather than larger ones, the tiny house movement offers intriguing insights. For instance, what lessons can a developer constructing standard semi-detached houses learn from tiny homes to shape the future of housing? Additionally, how can we transition from merely increasing project management to enhancing its quality? It may involve moving from traditional project management approaches towards more sustainable ones.

What is the game-changing aspect of the project you are currently working on or facing? When have you identified the trend that is most important to you and spent time understanding it. After a deep dive you will start to listen differently and direct the team's attention towards new insights.

Examples of trends in the construction industry are:

- From cheapest possible to as sustainable as we can imagine
- From standard contracts to partnering
- From blazing ahead to reflection and mindfulness
- From paper to digital

An essential tool for comprehending trends lies in recognising the various types that exist. By mapping them out for a project, you gain a deeper understanding of the landscape you're navigating.

Hard and soft trends: Hard trends are those that individuals cannot opt against adopting. For example, climate change is inevitable for both individuals and companies, making it wise to address. In contrast, soft trends are events that are undeniably occurring, with many people adapting accordingly. However, you or your project can choose, to a certain extent, whether to embrace them or not. An instance of this is the increasing preference for remote work post-pandemic. In your project, you may decide to accept this trend and benefit from it or reject it by requiring everyone to work in the office daily. Both paths have consequences, but unlike hard trends, soft trends offer a choice.

Megatrends are significant trends that greatly influence many aspects of life. One such megatrend is urbanisation. It involves a massive influx of people moving to cities, profoundly affecting the requirements for water and wastewater services in megacities. When combined with the trend of higher standards of living the need for water services may expand quite rapidly.

Anti-trends: Often, there is a counter-reaction to popular trends. Virtual reality (VR) allows individuals to wear a VR headset and explore planned architectural and civil engineering structures in a simulated environment. However, a contrasting trend involves creating physical models of buildings to facilitate a different type of conversation about the construction. Thus, although VR represents one trend, the opposite movement also gains traction.

Cyclical trends refer to recurring patterns that vary but maintain certain similarities. The preferred form of procurement could be considered a cyclical trend. For instance, the choice between design–bid–build or turnkey contracts for work execution is somewhat cyclical in nature. Each method has its challenges, and when they become too overwhelming, there may be a shift to the alternative procurement form. However, with each iteration, there is typically some improvement in the approach, whether it involves incorporating agile methodologies, partnering, win–win mind-set, or sustainability.

Exponential trends: Exponential trends start out very slowly and then accelerate rapidly. Artificial intelligence follows this pattern. In the beginning it had limited and nerdy applications, but within a very short time it is revolutionising entire professions.

Wildcards: A wildcard is an unlikely trend that, if realised, would have a significant impact. The growing number of people working independently or in small groups on consulting work could dramatically affect large consulting businesses, which would no longer hold such defining power in water. Instead, innovative and skilled individuals would collaborate to create entirely new types of projects. Conversely, this shift may not occur at all.

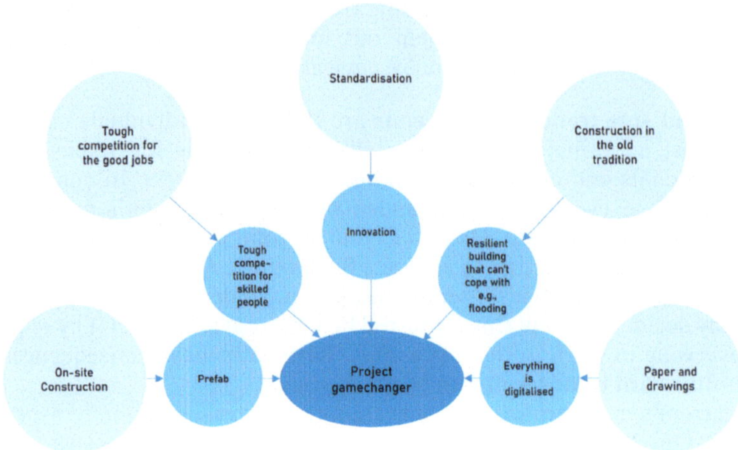

Figure 29 Game changer map for a project in the construction sector.

There are numerous ways to engage with trends. Collaborating on them allows us to develop a comprehensive map of the future, considering both visible trends and those we intuitively sense. This approach benefits from our unique perspectives and backgrounds. Clients, consultants, contractors, and subcontractors are likely to have differing views on which trends are apparent. By merging this knowledge and examining it as powerful ocean currents, we establish an intriguing foundation for devising visionary projects that either reshape the world or adapt to the evolving world we're entering.

One method to demonstrate the outcome is through a trend map that highlights the team's most significant trends for the project – the game changers. For instance, a trend map tailored for the construction industry is shown in Figure 29.

Trends and game changers are valuable tools for envisioning and communicating future prospects. By adopting a futurist perspective, the project can generate innovative ideas for collaboration, product design, stakeholder engagement, and documentation. The key lies in identifying the game changer(s) that will bring positive and novel advancements to the project.

Different team members may have different favourite game changers, but as long as you're working in the same general direction, there is ample room for varied experiences. There's no need to squeeze everyone into the same picture. But often, a team will find that when they put their images together, they can fit together and may even complement each other in a good way.

An approach to creating a better new world is to look at from A to B trends as journeys – one step at a time. Creating and using a trend map does not mean that everything in the project must be 'futuristic' in every detail. Often, just clarifying the future the project is aiming for is enough for the project to take both big and small steps in the right direction. The future is not created in a single project, but all projects should take small steps forward. If one hasn't considered the future, decisions will most likely be made based on past experiences. When we want to shift gear, it's crucial to explore and discuss the future.

THEORY U

Theory U is a leadership and facilitation philosophy that focuses on the quality of people's attention and the consequential ability to create change. Developed by Otto Scharmer from MIT (Scharmer, 2013), the theory incorporates ideas from various researchers and philosophers in the fields of consciousness and attention. It includes contributions from, for example, Humberto Maturana, a Chilean biologist, and philosopher; Peter Senge, a systems theorist and organisational development expert; and Edgar Schein, an organisational psychologist.

The U-process theory creates a framework for a process for groups based on being present – Otto Scharmer calls this 'presencing'. The word 'presencing' is derived from the words 'presence' and 'sensing', it involves tuning into one's inner knowing and intuition to bring about transformative change in leadership and organisational contexts.

The name of the theory comes from the shape of the change process, which is shaped like a U. In a theory U process, you start at the top-left corner of the U and work your way down towards the bottom of the U. After a reflective pause at the bottom of the U; you move up the right side of the U again. Along the way, there are a number of stations, as illustrated in Figure 30.

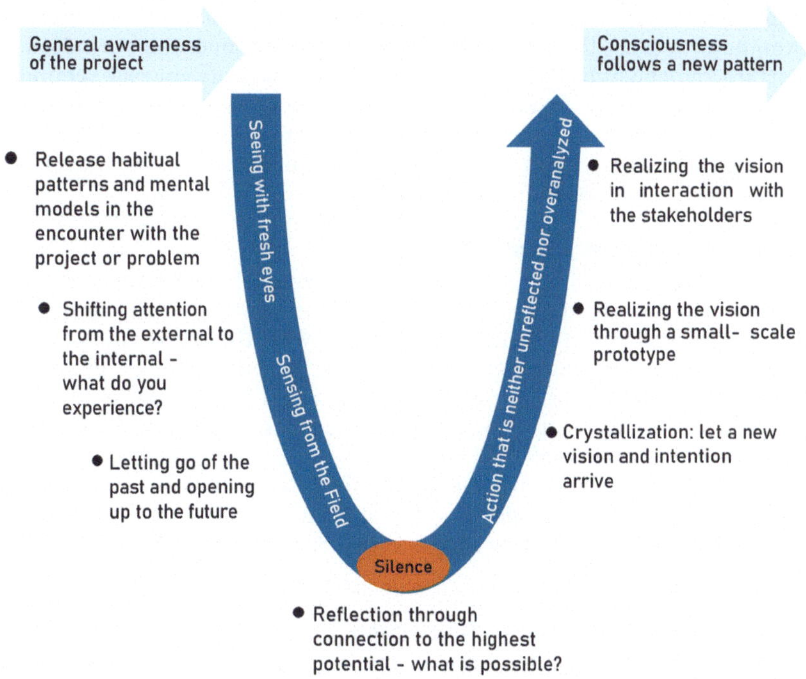

Figure 30 Process of theory U according to Otto Charmer (*source*: adapted by the author from Scharmer, 2013).

Left side process of the U: observe, observe, observe

The objective of the left side of the U is to attain a deep understanding of the issue that the group is exploring. The fundamental concept is for the group to 'learn from the future' rather than relying primarily on past experiences. This is not to undermine the value of experience, but rather an effort to approach the problem as if it were completely new, allowing for new possibilities. The three stages on the left-hand side focus on understanding the issue using an open mind, open heart, and open will.

The initial step involves setting aside habitual patterns and mental models to observe with an open mind from a fresh perspective. Scharmer advises 'examining reality from the edge of the system'. Often, we perceive things through the echo chamber we are part of; our insights are influenced by an invisible group thinking process involving those around us. Therefore, the goal is to discover new observation points and examine the selected issue anew, free from the preconceptions of what we believe we already know.

In the next step, we shift our focus from the object under examination to our own inner responses and felt sense of existence. We see ourselves within the system, recognising that we are part of it, cultivating an 'ecosystem awareness'. This transition occurs when we move from 'me observing' to 'me being included in the system'.

The third step involves subtly letting go, releasing our mental grip on our understanding to make room for new insights. This creates an inner space for fresh ideas to emerge in our imagination. We let go without anticipating what comes next.

Another way to see this process is: (1) we let go of our inner voice of judgement to nurture an open mind. (2) We release the inner voice of cynicism, reminding ourselves of limitations, to nurture space for an open heart. (3) Finally, we let go of fear and embrace an open will, or readiness to act.

During this process, groups often repeatedly redefine and refine the problem along the way as understanding deepens.

Bottom of the U: withdrawal and reflection

After the groups have gained a deeper understanding of the project or problem, they take part in a meditative retreat, ideally in solitude or in nature. This retreat helps individuals connect more deeply with themselves and consider the resonance of the process so far reflecting from the standpoint of 'their higher selves'. They may ask questions like 'Who am I?' and 'What is my work?' Following the retreat, the group gathers to share their reflections, followed by further individual contemplation.

The group eventually reconvenes to distil the essence of the insights that have emerged. Ultimately, they conduct group sharing sessions in a respectful and focused atmosphere, where everyone is conscious of 'honouring the important things that are happening'.

Right side of the U: translating into action

The right side is a co-creation phase where the group works on prototypes for the emerging future. The prototype should only be about 80% ready before it is presented to relevant stakeholders.

A key principle for the prototype is that it should be relevant, revolutionary, quick, and rough. Throughout the process, a balance should be struck to avoid the dangers of acting too recklessly fast on the one hand – and to avoid acting too slowly and potentially falling into 'analysis paralysis' on the other. Once there are one or a few prototypes, they should be presented to stakeholders for feedback. The feedback is used to adapt or modify the prototypes to make them better suited to the target audience. This can be realised through a number of feedback loops.

Then, the aim is to scale up the prototype so that it becomes 'alive' in the real world. The team will now be able to act from a shared awareness of the goal and all the details relevant to the 'product'. As work on the prototype progresses, ongoing capacity building within the group takes place.

Tools and tools

The Presencing Institute has a number of tools available online (www.u-school.org, 2023), including:

- Container building: describes the process of creating a safe space to work in.
- The iceberg model: a tool that makes underlying assumptions visible.
- Stakeholder interviews: an interview template to get a deep understanding with the
- stakeholders.
- Dialogue interviews: a recipe for reflexive and creative dialogue.
- Shadowing: method of observing stakeholders to gain understanding.
- Learning journey: recipe for travelling into the world of stakeholders and coming back with new knowledge.
- Case clinic: method to find the 'next step' in the creative process of finding innovative solutions.
- Three-dimensional modelling: group process where the group can see the problem unfold and experiment with the unfolding.
- Empathy walk: method to create a safe space and listen to someone else.
- Guided journalling: process of self-reflection questions.
- Story sharing: a way for participants to share stories from a perception of themselves.
- Seeds: helps to manifest ideas from what participants sense at the bottom of the U.
- Prototyping: creating a concrete prototype from an idea.
- Listening assessment: a daily evaluation method of your active listening skills.

Fault lines

Theory U focuses on understanding sustainability, by emphasising three specific gaps or fault lines that contribute to the creation of an unhealthy, unsustainable society. These three fault lines are:

(1) Our relationship with nature and our planet
(2) Our relationship with each other
(3) Our relationship with ourselves

Fault line 1: the ecological divide in 'our relationship with nature and our planet' is evident in various ways, such as increasing pressure on water supplies, water pollution, loss of topsoil, climate change, degradation of eco-habitats, and irreversible species loss. This divide has contributed to unprecedented environmental degradation and a decline in nature. A startling statistics reveals that we consume more than one and a half times the Earth's renewable resources each year.

Fault line 2: 'our relationship with each other' is a socio-economic divide in the gap between the haves and have-nots. This has led to unprecedented levels of inequality and fragmentation, resulting in a loss of sense of community, responsibility, and solidarity – fewer than 100 billionaires possess wealth equivalent to half of what humanity collectively owns.

Fault line 3: 'our relationship with ourselves' is a kind of spiritual divide and can be seen in rising rates of suicide, depression, and burnout. It is a divide between the self and the Self (capital S) – a kind of loss of 'spirituality' and consequent loss of 'meaning in life'. Every 40 s, someone commits suicide. More than 800,000 people commit suicide per year. This number surpasses the combined total of those killed by war, murder, and natural disasters.

Theory U posits that the three divides are interconnected. There are feedback loops between fault lines that amplify the vulnerability of our global society's collective fabric. We all play a part in creating these fault lines, and each day, the patterns are perpetuated – as if by an army of zombies trapped in an unshakable loop of collective history.

Otto Scharmer characterises this 'zombie state' with the concept of 'absencing' as the opposite of 'presencing'. People often navigate challenging situations on autopilot, with their consciousness disengaged. Humans tend to 'switch off' when confronted with difficult, unsafe, or painful experiences. Instead of addressing these issues, the automatic response is to disregard, avoid, or remain passive. The process of absence unfolds as follows:

(1) 'Denial' – not realising what is going on.
(2) 'Insensitivity' – that is, reducing emotional connection and thus cutting ourselves off from empathising with others.
(3) 'Absence' – that is, losing touch with one's higher self, ideals, and values.
(4) 'Delusional' – being under the influence of illusions and lies.
(5) 'Destruction' – where we harm ourselves and others. We resort to rage instead of patience.

This leads to a kind of tyranny characterised by an us-versus-them mindset – a rigid form of collectivism. Scharmer further explains this as a closing of the mind, heart, and will. When we address difficult situations from a mindset of absence, we place ourselves in a landscape where the risk of earthquakes along fault lines increases. Consequently, we directly and indirectly create new challenging situations.

The central question that theory U tries to answer is: how do we stop creating outcomes that nobody wants?

MAIN POINTS

Co-creation as a fusion of innovation and cooperation involves groups working collaboratively to generate novel outcomes. This chapter examines various methods for facilitating and participating in co-creation processes. As the world becomes more complex, co-creation will play a crucial role in discovering truly sustainable solutions.

The processes are presented according to increasing demands on facilitation skills.

Meeting management and facilitation involve working with groups in a manner like our current approach to cooperation. However, certain principles can help enhance the effectiveness of meetings. These principles encompass how project managers organise and skilfully guide participants through well-designed programmes. By adhering to these guidelines, meetings can yield better results and better experiences for all participants.

Mediation is a structured approach to decision making, primarily designed as a conflict management tool. It involves conflicting parties engaging in mutual understanding, identifying interests and needs, brainstorming solutions, and collaboratively deciding on a course of action. The technique can be applied in various decision making and design processes beyond conflict resolution. In both conflicts and other matters, mediation serves as an effective co-creation method.

Working with amateur 'future science' can be an enjoyable and fulfilling approach to future-oriented tasks, in contrast to working from a historical perspective. In their book 'Awakening your Future Sense' (Hovgaard, 2020), Skare and Hovgaard have made training your sense of the future into easily accessible tools for all. These tools can be remarkably effective in various projects and organisations.

Theory U is a method focused on fostering sustainable change management and co-creation. The process derives its name from the U shape, which begins in the top-left corner and encompasses various approaches to viewing an issue from a fresh perspective. At the base of the U, personal and collective reflection takes place. This leads to prototyping processes, culminating in the top-right corner where tangible change is implemented.

It is advisable to train and explore co-creation methods as soon as a group is ready. By familiarising yourself with the essentials of effective co-creation processes before the group's readiness, you can plant seeds and cultivate an environment that fosters the necessary culture for co-creation methods to work.

YOUR REFLECTIONS

Facilitating and participating in co-creation processes is next stage for a successful project manager. Luckily, numerous proven methods exist. Reflect on which principles are helpful in your situation and identify the competencies you need to develop to enhance your performance.

How do you think the group scores (0–10) on the following qualities:

- Inclusivity (embracing diversity)
- Clear purpose (expressed as needs or as stakeholder understanding)

- Psychological safety (so that everyone can contribute from their best place)
- Creativity (do you succeed in creating creative solutions)
- Responsibility for each other (relationships and getting the best out of each other)

What meeting management optimisations can you implement right away, and is there potential to utilise facilitation more effectively in project meetings? Perhaps, someone in the group excels at facilitation – it doesn't necessarily have to be you who takes on that role.

Experiment with mediation on either a non-conflictual issue or a hypothetical scenario. Be sure to follow the steps accurately. The main challenge arises when progressing through the stages too hastily. Take time to comprehend the issue fully. Here is an example for you to practice in a group.

Practising mediation

Here is a proposed scenario for training mediation. The mediation involves three participants, who receive individual instructions (not shared). Participants are encouraged to expand on the problem as needed.

Peter

As the project leader, you must submit a budget for the upcoming year, necessitating cost estimates from each of your five project participants. You communicated this during a project meeting a week ago, setting a deadline of 1st June. By 2nd June, George still has not provided his estimate.

Upon enquiry, George acts as if he is unaware of the request and denies ever having heard about it, though he was present at the meeting and the request is documented in the meeting minutes. This is not the first time George has failed to deliver on commitments, causing your frustration to mount.

When confronted, George dismissively claimed he wouldn't have time to prepare the estimate until next week – a completely unacceptable timeline given that you have only 2 days left to finalise the overall budget.

George

You are currently involved in three different projects. At the moment, project B demands the majority of your time and attention, due to its high pressure.

Earlier today, Peter the project manager of project A unexpectedly approached you and asked for budget figures. Upon expressing your surprise about the sudden request, Peter became angry, irritable, and his tone was quite frustrating – especially considering your stress with focusing on project B.

Politely, you informed him that you would have the necessary figures prepared by early next week.

Emma

As a mediator in the department, your role is to assist disputing parties in reducing conflict within the workplace. Today, Peter has approached you, expressing frustration with George. Peter believes George is not fulfilling his responsibilities on Peter's project as previously agreed upon. The issue seems to involve a budget, but further details remain unclear.

Your objective is to guide both parties through the five stages of mediation. Start by gathering a free narrative from each participant. Next, facilitate the identification of their individual needs. Utilise the mediation process by having them work together to brainstorm methods of addressing each other's needs. Ultimately, both parties will agree on specific solutions and determine the action steps required for moving forward.

Futurist exercise

Individually or in groups, attempt to identify trends as shifts from A to B. Then, work on recognising trends that significantly impact your current project. Can you distinguish between the types of trends: hard and soft trends, megatrends, anti-trends, cyclical trends, exponential trends, and wildcards? What are the key game changers for you and your project? Dedicate some time to interpreting news as trends – determine what each news item represents and how you would describe the corresponding trend. Eventually, it becomes similar to analysing meteorological maps with varying highs, lows, and storm patterns. By recognising compelling trends, numerous new opportunities arise.

Consider exploring theory U when you feel prepared. Initially, you might want to carry out a theory U process with yourself or a friend before applying it to larger groups. Theory U is an effective method for fostering genuine change and excels at elucidating the internal mechanisms that enable change. The Presencing Institute offers regular online workshops, which may prove beneficial as well.

READ MORE

We learnt mediation from Tina Monberg, and we recommend her books and courses to learn the subject. The book 'Handbook of Human Conflict Technology' goes through the mediation process and also includes how to set-up conflict strategies and so on (Monberg, 2019).

Anne Skare Nielsen and Henrik Good Hovgaard have published a number of books, of which 'Awaken Your Future Sense' is particularly central (Nielsen & Hovgaard, 2020).

There are many sources for understanding theory U. Of the international sources, Otto Scharmer's books and the Presencing website (www.u-school.org, 2023) are the most basic. 'The Essentials of Theory U: Core Principles and Applications' (2018) and 'Theory U: Leading from the Future as It Emerges' (Scharmer, 2013) are the most recent and most up-to-date in terms of evolving knowledge and understanding.

doi: 10.2166/9781789063349_0239

The Way Forward

If you want truly to understand something, try to change it.

Kurt Lewin

Kurt Lewin's quote reveals a fundamental truth: we can only understand something deeply when we work on changing it. This is evident when we try to change ourselves, realising the immense difficulty involved. Success in leading ourselves through transformations, such as transitioning from an 'ordinary project manager' to a 'sustainable project manager', hinges on our ability to alter our thought patterns and behaviours. Similar transformations include shifting from 'ordinary projects' to 'sustainable projects' and evolving from 'our society today' to 'the sustainable society of the future'. This change does not result from a few abrupt, dramatic shifts. It doesn't hinge on just thinking about change or complaining that the change has not already come about. Rather change requires a lot of small, incremental changes and adjustments. A steady, deliberate, reflected approach is the key to success.

If we are not careful the idea of sustainability risks becomes diluted, greenwashed, and superficial – signifying everything and nothing simultaneously, a false halo that offers no benefits. Large-scale systems worldwide must undergo change in a coordinated manner. On the one hand, we can observe that some of these changes are already happening; consider the progress made over the past 10, 20, or 30 years. However, on the other, we must also acknowledge that some ideas and trends continue to lead us astray on various fronts. Our individual responsibility is to identify where we can contribute to changing the direction of these systems. Simultaneously, we must trust that similar agents are doing the same in different systems and that together it amounts to real change. We have to take sustainability seriously without being overwhelmed by its gravity.

UNDERSTANDING CHANGE BIOLOGICALLY

Human beings possess immense potential. Our bodies and minds are brimming with capabilities we have yet to discover. It's comparable to a car equipped with countless features such as reverse gear, lane-keeping assist, autopilot, and thousands of other options. There are numerous buttons we haven't even pressed or realised their existence. Likewise, there are countless possibilities around us that remain unexplored only because we haven't yet dared or imagined them.

As we focus more on realising our potential, we unlock an increasing number of capabilities, allowing us to handle various situations with greater ease. We can develop new skills or extend existing ones, but to do so, we must switch off our inner autopilot and stay conscious of our actions. By being aware of the evolving aspects of ourselves, we never have to feel bored. There will always be changes to undertake – all sorts of small self-defined projects and 'developments'. Additionally, we can continually reflect on what works and what doesn't, then ask ourselves: 'So, what would be a wise approach to try tomorrow?'

The task of realising our potential as human beings in the role of project managers closely parallels the challenge and difficulty of shifting the world towards a more sustainable direction. In this regard, self-improvement offers actual experience where we can learn by doing. Changing ourselves help us comprehend system change at larger scales.

Given our serious and complex sustainability challenges, the classic (and a bit tired) mantra of 'wanting to make a difference' is only the beginning. If you are a person who wants to make a difference, the first stumbling block is that you have unclear expectations and goals. Your unclear expectations risk being both too high and too low at the same time. To set good expectations: create experiments. Experiments are the best way to learn to understand what can be achieved. The trials offer valuable lessons in determining viable paths to success and can be scaled up as capability grows. Most importantly, experiments provide learning opportunities regardless of their outcomes. Great outcome – perfect. Failed outcome – great, what can we learn?

Three types of unclear expectations, in particular, can create unintended outcomes that, over time, can prove counterproductive to the desire to move towards greater sustainability. The problems can be categorised into:

(1) Expect that you as an individual can bring about rapid change.
(2) Believing that you with your direct actions, create one-to-one changes.
(3) Focusing too much on seeing clear results as direct results of your own work.

The root of all three problems lies in our fragmented perception of the world and our insufficient understanding of the systems we engage with. It is crucial to consider biology as a good model for understanding what change is and how it comes about. We ought to change our thinking into thinking of systems as living beings rather than as machines that can be manipulated. It is helpful to recognise the biological aspect of our society at every level: our social circle, the organisations we work for and collaborate with, our nation as a unified entity, global humanity, and the overarching biological systems that encompass all living beings on Earth. We must be aware that these systems are interconnected and interdependent – 'everything is connected'.

When we visualise these systems, we understand that we are very small in the big systems. At the same time, through 'the great interconnectedness', we are big, as something we have said or done can make an impression on someone else, who consciously or unconsciously copies the behaviour with or without our being aware of it. Our resulting change may happen far away from us and much later and we may never know about it. You as well may be one of many agents spreading the behaviour from something you have seen. You may not be the leader of the new – an equally important role is to follow and copy what works. When we change, we are learning about and affecting systems at the same time. Many more good changes may occur if we don't depend on taking the credit for the change.

Living systems operate on different principles than the machine metaphor. Still unconsciously the machine metaphor imperceptibly informs our thinking in so many ways. The frustration is that 'the buttons don't work properly'. That is to say, my system and the system I am a part of do not do what I intend. Quite simply, if my intention is to lose weight, it is far from certain that is what happens – or if it does happen, that it is sustained in the long run and actually leads to me becoming healthier; just as it is obvious to everyone that I cannot decide on my goal today and achieve it tomorrow. If I want to go from being short-haired to long-haired, from thin to thick, or from a couch potato to an Olympic athlete, it takes time, reflection, creativity, focus, patience, and will.

Comprehending oneself as a biological entity or as a collection of systems within a larger framework is essential for devising good experiments. The idea of living systems – biota – serves as an ideal foundation when working towards sustainability. It possesses very different attributes that our unconsciously favoured metaphor, 'the machine'. Contemplate the following four questions:

- How can I change myself as a living system?
- How can I change the world around me through projects – thinking of it as a living system?
- How can I change the way we work together – thinking of us/the group as a living system?
- How can I change my project management competences in the direction of living systems?

Attempt to answer the questions using both a living system and a machine metaphor. The two approaches are likely to yield quite different possibilities and methods.

Understanding living systems and recognising the slowness and interconnectedness of natural processes reveals why the abrupt changes we sometimes tend to impose on ourselves or others are ineffective. They are as unsustainable as a whiplash injury to the neck.

For lasting change, it's essential to place biological understanding at the core of our thoughts and actions. This way, ideas about movement and change can be expressed through small steps that gently support our current state of progress.

Envision a sustainable future in 2030 – what values, indicators, behaviours, and other elements do you believe will be present? How can you contribute to the collective goal by reinforcing a particular flow? How will it look if built on principles of living systems thinking?

Change management introduces uncertainty into an organisation. When transitioning from the 'old way' to a 'new way' of doing things many individuals will feel uneasy or uncertain about adopting new behaviours. This is primarily because mastering new competencies involves a period where you may feel incompetent, while learning new routines and unlearning old ones. This experience can be unsettling and requires a return to a learning mindset – embracing the learning focus again and again.

Psychologist Kurt Lewin (1890–1947) observed that both individuals and organisations tend to maintain the status quo, resist change, and preserve existing conditions. However, we face constant changes in our daily lives, many of which we accept unconsciously. One of Lewin's renowned models is 'unfreeze–change–refreeze'. This model suggests a three-step process for executing planned changes. The first step, 'unfreeze', involves challenging current behaviour and raising awareness about the necessity for change. The second step encompasses the change itself. Finally, in the third step, the change is solidified to prevent reverting to old behaviours. As a change agent, it's crucial to remember all three steps for success.

The success of projects depends very much on the inner orientation of the project manager. Therefore, although project management tools are important to apply systematically and disciplined, they will only work if you, as a project manager, invest yourself in the project. This necessitates a solid understanding of one's own values, living in accordance with those values, and maintaining ethical clarity. Rather than being a mere mental exercise, this is a practical, ongoing endeavour where you as the project manager continuously improve your ability to work in alignment with values and ethics relevant to the project's purpose, participants, customers, stakeholders, and larger context and tackle the problems and issues in and around the projects creatively and constructively.

Throughout your journey, you'll undoubtedly face emotional challenges. Leadership can be frustrating, and setbacks are inevitable. Misunderstandings and antagonism will arise at times – take it as a learning experience.

As a project manager acting as a change agent, you have a unique opportunity to make a significant impact with your projects. It's worth noting that many people are more receptive to their peers than to management messages during a change process. You have as a project manager to act as a change agent and the potential to drive a movement towards a sustainable future.

If everyone waits for someone else to initiate change, progress will never happen. It's essential for someone to take the first step where they are – will that person be you?

Thank you

We have conducted hundreds of experiments to create superior projects, enriched by individuals who inspired us. We would like to extend special thanks to Tina Monberg, who brought us together years ago and encouraged us to adopt a leadership style focused on a positive, peaceful, and loving future. Additionally, Gustaf Olsson, Michael Stubberup, and Steen Hildebrandt have been significant sources of inspiration.

Valuable feedback and assistance were provided by multiple reviewers, including Jens Kudal Borchmann and Anders Bjarke Thomsen from Hillerød Forsyning, Karina Buus from Envidan, Gitte Godsk Dalgaard from COWI, Michael Ib Thomsen from NCC, Trine Bøgelund-Kjær and Anne Støtt Hansen from Winsløw, and Ole Steen Andersen from Factor3. Moreover, Pernille Lethenborg, Anne-Grethe Ingildsen, and Mette Nielsen diligently reviewed the text, ensuring grammatical accuracy and comprehension. During the translation process we have received great help from Frida Ingildsen and Ejner Thorkilsen. We are proud that our grown-up children have been able to help us in this endeavour.

We also express our gratitude to those we collaborated with on various projects such as The Øresund Fixed Link, DR, the Affordable Housing Project in Copenhagen, Psychiatric Centre Sct. Hans in Roskilde, as well as project managers at Prevas, Kalundborg Forsyning, Hillerød Forsyning, and all the other major and minor projects we participated in.

We would also like to extend our thanks to Mette Berthelsen at Molio for making this project possible and offering unwavering support throughout the process. And finally, we want to thank the team at International Water Association Publishing, especially Rod Cookson and Andrew Peart.

References

Agilemanifesto (2023). Manifest for Agile Software Development. https://agilemanifesto.org.

Alexander K. N. (2017). Living in a Mindful Universe. Rodale Books, Emmaus, Pennsylvania.

Aurelius M. (2023). Meditations. Reader's Library Classics, Delhi, India.

Beck C. C. (2014). Spiral Dynamics: Mastering Values, Leadership and Change. Wiley, Hoboken, New Jersey.

Belbin R. M. (2010). Team Roles at Work. Taylor & Francis Ltd, London, UK.

Benyus J. M. (2002). Biomimicry – Innovation Inspired by Nature. William Morrow & Company, New York, USA.

Blincoe K. (2022). Achieving Sustainability: The Ultimate Human Challenges: Critical Barriers and Future Perspectives. Springer International Publishing AG, New York, USA.

Brundtland G. H. (1987). Our Common Future: Report of the World Commission on Environment and Development. Oxford University Press, Oxford, UK.

Changali S., Mohammad A. and van Nieuwland M. (2015). The Construction Productivity Imperative. McKinsey Productivity Sciences Center, New York, USA.

Childre D. and Martin H. (1999). The HeartMath Solution: The Institute of HeartMath Revolutionary Program of Engaging the Power of Heart's Intelligence. HarperCollins Publishers, New York, USA.

Covey S. R. (2020). The 7 Habits of Highly Effective People – 30th Anniversary Edition. Powerful Lessons in Personal Changes. Simon + Schuster Inc, New York, USA.

Cox D. R. (2012). SCARF. Updating the social neuroscience of collaborating with others. *NeuroLeadership Journal*, USA.

Crawford J. K. (2021). Project Management Maturity Model. Taylor & Francis Group, New York, USA.

Daniel Goleman R. E. (2016). Primal Leadership – Unleashing the Power of Emotional Intelligence. Harvard Business School Press, Boston, USA.

David H. and Maister R. G. (2021). The Trusted Advisor. Free Press, Florence, MA.

Dispenza J. (2014). You are the Placebo – Making Your Mind Matter. Hay House, Carlsbad, California.

Dispenza J. (2019). Becoming Supernatural. Hay House, Carlsbad, California.

Doughnuteconomics (24 September 2023). Doughnuteconomics. Retrieved from https://doughnuteconomics.org/

Earle S. E. (2010). The World is Blue. National Geographic Society, Washington, USA.
ECI (2023). www.eci.org. European Construction Institute.
Epictetus (4 February 2023). Stoic Rules of Life. Taken from Peoples Press, Copenhagen, Denmark.
Evans C. S. (2009). Kierkegaard: An Introduction. Cambridge University Press, Cambridge, UK.
Ferguson R. (2013). Life Lessons from Kierkegaard. Pan Macmillan, Hampshire, UK.
Fuller R. B. (1969, 2008). Operating Manual for Spaceship Earth. Lars Muller Publishers, Zurich, Switzerland.
Gaarder J. (2003). Sophies World. Orion Publishing Co, London, UK.
Gaffney J. R. (2021). Breaking Boundaries: The Science of our Planet. Dorling Kindersley Ltd, London, UK.
Giordano P. (2020). How Contagion Works: Science, Awareness and Community in Times of Global Crises – The Short Essay that Helped Change the COVID-19 Debate. Orion Publishing Group Limited, London, UK.
Goleman D. (2005). Emotional Intelligence. Bantam Trade, New York, USA.
Goleman R. D. (2018). The Science of Meditation: How to Change Your Brain, Mind and Body. Penguin Books Ltd, London, UK.
Greenleaf R. K. (2002). Servant Leadership. Paulist Press, New Jersey, USA.
Hanselman R. H. (2016). The Daily Stoic: 366 Meditations on Wisdom, Perseverance, and the Art of Living. Portfolio, New York, USA.
Hanselman R. H. (2022). Lives of the Stoic: The Art of Living from Zeno to Marcus Aurelius. Profile Books Ltd, London, UK.
Hicks D. (2021). Dignity: Its Essential Role in Resolving Conflict (10th Anniversary Edition). Yale University Press, Connecticut, USA.
Hildebrandt S. and Stubberup M. (2016). Sustainable Leadership: Leadership from the Heart. Copenhagen Press, Copenhagen, Denmark.
Hildebrandt S. (2018). Verdensmålene – det vigtigste punkt på dagsorden (The SDG – the Most Important Topic at the Agenda). KLS Pureprint, Copenhagen, Denmark.
Holiday R. (2014). The Obstacle is the Way: The Ancient Art of Turning Adversity to Advantage. Penguin Publishing Group, London, UK.
https://doughnuteconomics.org/Creating-City-Portraits-Methodology.pdf (16 September 2023). Retrieved from www.doughnuteconomics.org
https://www.hearthmath.org (24 September 2023). Retrieved from HeartMath Institute: https://www.heartmath.org/
Ingildsen P. (2020). Water Stewardship. IWA Publishing, London, UK.
Inner Development Goals I. (2022). Inner Development Goals: Transformation Skills Sustainable Development. Retrieved from www.innerdevelopmentgoals.org
IPMA (2022a). Competence in Project Management. Assessment Tool for Managing Project, Programs and Portfolios. International Project Management Association, IPMA, The Netherlands.
IPMA (24 September 2022b). Retrieved from www.ipma.world
Jain N. (2016). The Most Scarce Resource on the Planet: Mindset of Abundance. Retrieved from www.youtube.com/watch?v=KX78jynWLD0
Kahneman D. (2012). Thinking, Fast and Slow. Penguin Books Ltd, London, UK.
Ken Wilber T. P. (2008). Integral Life Practice. Shambhala Publications Inc, Boulder, Colorado, USA.
Kimmerer R. W. (2020). Breading Sweetgrass, Minneapolis, USA.
Kolbert E. (2014). The Sixth Extinction, New York, USA.
LCI (2023). https://leanconstruction.org/. LCI.

Leeds R. (2017). Top 4 Challenges Facing the Construction Industry. https://www.futureofconstruction.org.
Liker J. (2021). The Toyota Way. McGraw-Hill Education, USA.
Lipton B. H. (2015). The Biological of Belief: Unleashing the Power of Consciousness, Matter & Miracles. Hay House UK, London, UK.
Lyngby A. J. (1999). The Environmental Management and monitoring of the Øresund Fixed Link. Terra et Aqua, Number 74, The Netherlands.
Maslow A. (2012). A Theory of Human Motivation. Lushena Books, Chicago, USA.
Monberg T. (2019). Handbook of Human Conflict Technology. Paragon Publishing, Rothersthorpe, Northampton, UK.
Morton T. (2013). Hyperobjects: Philosophy and Ecology after the End of the World. University of Minnesota Press, Minnesota, USA.
Nash J. F. (1950). Nash Equilibrium, Massachusetts, USA.
Von Neumann J. and Morgenstern O. (2004). *Theory of Games and Economic Behavior.* 60th Anniversary Edition. Princeton University Press, Princeton, New Jersey, USA.
Nielsen A. S. and Hovgaard H. G. (2020). Awaken Your Future Sense. Independently Published, Copenhagen, Denmark.
Olsson J. R., Ahrengot N. and Attrup M. L. (2019). Power in Projects, Programs and Portfolios. Djøf Publishing, Copenhagen, Denmark.
Peek S. (u.d.). Mary Parker Follett's Management Theory. https://www.business.com/articles/managementtheoryofmaryparkerfollett/
PMI P. M. (2021). A Guide to the Project Management Body of Knowledge. PMI, Pennsylvania, USA.
Raworth K. (2018). Doughnut Economy: Seven Ways to Think Like a 21st-Century Economist. Cornerstone, London, UK.
Richardson J., Steffen W., Lucht W., Bendtsen J., Cornell S. E. *et al.* (2023). Earth beyond six of nine Planetary Boundaries. *Science Advances*, **9**, 37.
Robinson K. S. (2021). The Ministry of the Future. Orbit, London, UK.
Rogers R. E. (2015). Active Listening. Martino Fine Books, Connecticut, USA.
Rockström J., Steffen W., Noone K., Persson Å. *et al.* (2009). A safe operating space for humanity. *Nature*, **461**, 472–475. DOI 10.1038/461472a
Rosenberg M. B. (2015). Non-Violent Communication. Puddle Dancer Press, Encinitas, CA, USA.
Sachs J. D. (2015). The Age of Sustainable Development. Columbia University Press, New York, USA.
Sanford C. (2022). Indirect Work: a Regenerative Change Theory for Businesses, Communities, Institutions and Humans. InterOctave, Washington, USA.
Sarhan S. A. (2013). Barriers to Implementing Lean Construction in the UK. Faculty of Science and Technology, University of Plymouth, Plymouth, UK.
Scharmer O. and Kaufer K. (2013). *Leading from the Emerging Future: From Ego-system to Eco-system Economics*. Berrett-Koehler Publishers Inc. Oakland, California, USA.
Scott B. (2000). Partnering in Europe – Incentive Based Alliancing for Projects (ECI). Thomas Telford Publishing, London, UK.
SCRUM (24 September 2023). Scrum. Retrieved from www.scrum.org.
Senge P. M. (1999). The Dance of Change: the Challenges to Sustaining Momentum in a Learning Organization. Broadway Business, Manchester, UK.
Seymour J. O. (1997). Introducing Neuro-Linguistic Programming: Psychological Skills for Understanding and Influencing People. HarperCollins Publishers, New York, USA.
Smith A. (1776). The Wealth of Nations, USA.

Søltoft P. (2018). 10 ting ledere kan lære af Kierkegaard (10 Things Leaders Can Learn from Kierkegaard). Akademisk Forlag, Copenhagen, Denmark.

Steffen W., Richardson K., Rockström J., Cornell S. E. *et al.* (2015). Planetary boundaries: Guiding human development on a changing planet. *Science*, **347**, 736, 1259855

Tamm R. J. (2005). Radical Collaboration: Five Essential Skills to Overcome Defensiveness and Build Successful Relationships. HarperCollins Publishers, New York, USA.

Taylor D. B. (2022). Whole Brain Living: The Anatomy of Choice and the Four Characters That Drive Our Life. Hay House, USA.

Thorkilsen C. D. (2001). An Owner's View of Hydroinformatics: Its Role in Realising the Bridge and Tunnel Connection Between Denmark and Sweden. *Journal of Hydroinformatics*, IWA Publishing 2001, London, UK.

Thunberg G. (2022). The Climate Book. Penguin Books Limited, London, UK.

Tolle E. (2004). The Power of Now: A Guide to Spiritual Enlightenment. New World Library, California, USA.

Tolle E. (2008). A New Earth: Awakening to Your Life's Purpose. Penguin LCC, New York, USA.

Tomm K. (2022). Circular Questioning Technique – The 4 Types of Questions. Retrieved from https://www.youtube.com/watch?v=AGiTJ-Z_y6Q&t=2s

Tuckmann B. W. (1965). Development sequence in small groups. *Psychological Bulletin*, **63**(6), 384–399, USA, https://doi.org/10.1037/h0022100

Ury R. F. (2011). Getting to Yes: Negotiating Agreement without Giving in. Penguin LCC, New York, USA.

Vernadsky V. I. (2013). The Biosphere. Copernicus Books, Germany.

Weisman A. (2022). The World Without Us. Ebury Publishing, London, UK.

Wohlleben P. (2018). The Hidden Life of Trees. Greystone Books, Canada.

Wright S. (2014). First Intelligence: Using the Science & Spirit of Intuition. New World Library, California, USA.

www.science.org (18 September 2023). Retrieved from https://www.science.org/doi/10.1126/sciadv.adh2458

www.unep.org (16 September 2023). Retrieved from https://www.unep.org/explore-topics/disasters-conflicts/where-we-work/sudan/what-integrated-water-resources-management

www.unwater.org (16 September 2023). Retrieved from https://www.unwater.org/our-work/integrated-monitoring-initiative-sdg-6/indicator-651-degree-integrated-water-resources

www.u-school.org (16 September 2023). Retrieved from https://www.u-school.org/resources

Index

E

F

G

H

I

R

S

T

U

V

W

IWA Publishing's authorised EU representative for General Product Safety Regulations is Diane D'Arras, 15 rue Duret, 75116 Paris, France, e-mail: safety@iwap.co.uk.

Printed and bound by CPI Group (UK) Ltd, Croydon, CR0 4YY
12/05/2026
02108869-0004